Excel

Get the Results You Wa

Year 8 Mathematics Practice Tests

Lyn Baker

PASCAL PRESS

Reprinted 2021, 2023, 2024

ISBN 978 1 74125 635 2

Pascal Press
PO Box 250
Glebe NSW 2037
(02) 9198 1748
www.pascalpress.com.au

Publisher: Vivienne Joannou
Project editor: Rosemary Peers
Edited by Rosemary Peers
Answers checked by Peter Little
Typeset by Kim Webber
Printed by Vivar Printing/Green Giant Press

Contents

NUMBER AND ALGEBRA

Mixed Measurement

Pythagoras' Theorem

Congruence and Properties of Quadrilaterals

STATISTICS AND PROBABILITY

Probability

Statistics

YEARLY EXAMINATIONS

Introduction

- The main purpose of this book is to provide students with **extensive test and exam practice** at all levels in Year 8, in the topics they cover in class. This way students will be given the test and exam practice they require to be fully prepared to excel at their class tests and exams in class.
- This book covers all the topics in **Year 8 Australian Curriculum Mathematics.**

The following features will help students excel in their class tests and exams:

- This book is a **workbook** that students can write in, ensuring that they have all their tests, exams and working in one place. This is invaluable when revising for topic tests and exams—no lost pages or notes.
- Each topic of work has a **pretest**. The pretests have been designed to check that students have the necessary skills for each section of work and this has determined their position in the book.
- The main **tests** all have **three sections**: Part A has multiple-choice questions; Part B has short-answer questions and Part C consists of questions that require some working out.
- The tests also have three different **levels of difficulty**.
  - **Level 1** tests are the easiest. They should determine whether a student has a basic understanding of the topic.
  - **Level 2** tests are aimed at the average student. Those who do well in these tests will have a solid grasp of the content of the topic.
  - **Level 3** tests are the most challenging. They are aimed at good students and should determine whether they have a thorough knowledge of the content and are able to apply that knowledge.

  Together these different tests can also be used as carefully **graded exercises** for revision.
- Some tests have been designed as **non-calculator tests**, allowing an examination of students' ability to work with whole numbers, fractions and decimals, without the use of a calculator. This is an integral part of the Year 8 course.
- Other tests in these topic areas are specifically **calculator allowed** to make sure students are also familiar with the working and limitations of calculators. It is assumed that students will have access to a calculator but in most cases the use of a calculator will not be essential.
- For some topics there are separate tests for the individual components of the curriculum as well as a **mixed test** that combines all those elements.
- There are **six yearly examination papers**. There is one at each level specifically designed to be done without a calculator, plus one at each level that is calculator allowed.
- **Answers** to all questions are provided at the back of the book.
- Note that the **innovative test and exam format** of this book can be used both as a means of assessing students' knowledge and understanding, and also to provide comprehensive revision of the course.

There are four titles in the series:

The Four Operations with Integers Non-calculator

PRETEST

Time allowed: 15 minutes **Total marks: 16**

Marks

1 Which of 0, –24, 17 and –36 is the smallest?

(A) 0 (B) –24 (C) 17 (D) –36 [1]

2 What number is found on the number line at the position of the arrow?

(A) –2 (B) –3 (C) –4 (D) –6 [1]

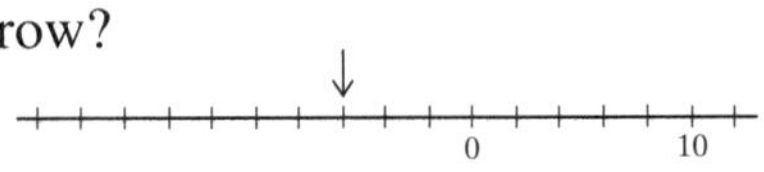

3 Which numbers are in ascending order?

(A) –3, 5, –9 (B) –9, 5, –3 (C) –9, –3, 5 (D) –3, –9, 5 [1]

4 The temperature was –3 °C. If the temperature fell 4°, what was the new temperature?

(A) –7 °C (B) –1 °C (C) 1 °C (D) 7 °C [1]

5 Which arrow points to –20?

(A) A (B) B (C) C (D) D [1]

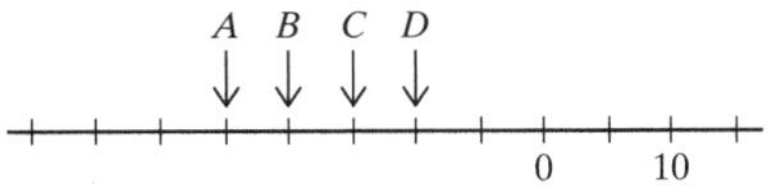

6 Which of the numbers –235, –372, –126 and –198 is the largest?

(A) –235 (B) –372 (C) –126 (D) –198 [1]

7 The temperature was –6 °C and is now 2 °C. The change is a:

(A) fall of 4° (B) fall of 8° (C) rise of 4° (D) rise of 8° [1]

8 Which is correct?

(A) –7 > –3 (B) 2 < –9 (C) –30 < –20 (D) –47 > –35 [1]

9 –2 + 3 =

(A) –1 (B) 1 (C) –5 (D) 5 [1]

10 5 – 8 =

(A) –13 (B) –3 (C) 3 (D) 13 [1]

11 –6 – 7 =

(A) –13 (B) –1 (C) 1 (D) 13 [1]

12 –9 + (–3) =

(A) –12 (B) –6 (C) 6 (D) 12 [1]

13 18 – –6 =

(A) –24 (B) –12 (C) 12 (D) 24 [1]

14 15 + (–5) =

(A) –20 (B) –10 (C) 10 (D) 20 [1]

15 –11 – –4 =

(A) –15 (B) –7 (C) 7 (D) 15 [1]

16 2 + 3 × 4 =

(A) 14 (B) 16 (C) 18 (D) 20 [1]

Total marks /16

LEVELS 1 AND 2 TEST — PART A

Time allowed: 40 minutes

Marks

1 $-2 \times -3 =$
(A) –6 (B) –5 (C) 5 (D) 6 [1]

2 $-54 \div 6 =$
(A) –9 (B) –8 (C) 8 (D) 9 [1]

3 $-3 \times -4 \times -5 =$
(A) –7 (B) –60 (C) 7 (D) 60 [1]

4 $(-3)^2 =$
(A) –6 (B) –9 (C) 6 (D) 9 [1]

5 $-1 \times -1 \times -1 \times -1 \times -1 =$
(A) –5 (B) –1 (C) 1 (D) 5 [1]

6 $10 - 2 \times 7 =$
(A) –4 (B) 4 (C) –24 (D) 56 [1]

7 $20 \div -2 \times 5 =$
(A) –50 (B) –2 (C) 2 (D) 50 [1]

8 $6 \times -4 \div -2 =$
(A) –12 (B) –26 (C) 26 (D) 12 [1]

9 $-3 \times 5 - 6 \times -1 =$
(A) –21 (B) –9 (C) –3 (D) 21 [1]

10 $-8 + 7 - 6 - 5 + 2 =$
(A) –12 (B) –2 (C) –10 (D) –8 [1]

11 $4 - 2 \times 6 + 5 =$
(A) 17 (B) 22 (C) –3 (D) –13 [1]

12 $-3 \times (4 - 5) =$
(A) 3 (B) –3 (C) –7 (D) –17 [1]

13 $\dfrac{8 + 6}{3 - 5} =$
(A) 5 (B) –7 (C) –5 (D) 7 [1]

14 $-18 \div -3 - 6 =$
(A) –12 (B) 6 (C) 0 (D) 2 [1]

15 $-5 - -4 \times -3 =$
(A) 7 (B) 3 (C) –3 (D) –17 [1]

The Four Operations with Integers Non-calculator **(Levels 1 and 2 Test)**

PART B Write answers only in the space provided.

16 $-5 \times 2 =$ ________________ 1

17 $-20 \div 4 =$ ________________ 1

18 $-3 \times -3 \times -2 =$ ________________ 1

19 $-4 \times 5 - 3 =$ ________________ 1

20 $7 - 6 \times 2 =$ ________________ 1

21 $30 \div 5 \times -2 =$ ________________ 1

22 $-8 + 12 \div -4 =$ ________________ 1

23 $-3 + 5 - 9 =$ ________________ 1

24 $(-4 - 3) \times (-2 - 8) =$ ________________ 1

25 $(-2)^3 =$ ________________ 1

PART C Show all working.

26 Given that $37 \times 48 = 1776$ what is the value of:

a -37×-48? 1 mark

____

b $1776 \div -37$? 1 mark

____

c $-1776 \div 48$? 1 mark

____

27 Find the value of:

a $(-7 - -8) \times (-9 + (-3)) \div (5 - 11)$ 2 marks

____

____

b $(16 - 18 \div -3 \times -2) \times (-5 + 3 \times -10 \div 5)$ 2 marks

____

____

c $-7 - \dfrac{-21 \div (3 - 4)}{9 \div -3 + 6}$ 2 marks

____

____

28 The maximum and minimum temperatures, in degrees Celsius, are shown for a city over three days.

	Sat.	Sun.	Mon.
Maximum	–4°	–2°	3°
Minimum	–17°	–15°	–10°

a What was the difference in maximum temperatures on Saturday and Sunday? 1 mark

____

b What was the difference in minimum temperatures on Saturday and Monday? 1 mark

____

c Which day had the biggest difference between maximum and minimum temperatures? 1 mark

____

d What was the average minimum temperature? 1 mark

____

e How much greater was the average maximum temperature than the average minimum temperature? 2 marks

____

____

Total marks /40

LEVELS 2 AND 3 TEST — PART A

Time allowed: 40 minutes

Marks

1 $-5 + (-6) - -8 =$

(A) -3 (B) 7 (C) 9 (D) 19 — 1

2 $-3 \times -2 - 5 =$

(A) 1 (B) -1 (C) -11 (D) 49 — 1

3 $8 - 16 \div -4 =$

(A) 2 (B) -2 (C) 4 (D) 12 — 1

4 $-32 \div -8 \div -2 =$

(A) 2 (B) 8 (C) -2 (D) -8 — 1

5 $7 + (-9) \times -5 =$

(A) 10 (B) -38 (C) 80 (D) 52 — 1

6 $-4 \times -6 - 3 \times -8 =$

(A) -48 (B) -288 (C) 48 (D) 0 — 1

7 $-36 \div -4 \times 3 =$

(A) 27 (B) 3 (C) -3 (D) -27 — 1

8 $12 - 7 \times -9 =$

(A) -45 (B) -51 (C) 45 (D) 75 — 1

9 $(-2)^4 =$

(A) 8 (B) -8 (C) 16 (D) -16 — 1

10 $15 - 27 + 8 - 56 - 17 =$

(A) -77 (B) -123 (C) -93 (D) 19 — 1

11 $-2 \times 3 \times 4 \times 0 \times -5 \times -6 =$

(A) -720 (B) 0 (C) 30 (D) 720 — 1

12 $-8 + 7 \times -3 - -4 =$

(A) 7 (B) -1 (C) 1 (D) -25 — 1

13 $14 - 2 \times 9 - 20 \div -2 =$

(A) 12 (B) 6 (C) 66 (D) -44 — 1

14 Which of these statements is correct? I: $-2 \times -3 \times -4 = 24$ II: $-3 \times -5 \times -2 \times -4 = 120$

(A) I but not II (B) II but not I (C) both I and II (D) neither I nor II — 1

15 $(-5 - 3) \times (-2 + 7) =$

(A) -72 (B) -40 (C) 40 (D) 72 — 1

PART B Write answers only in the space provided.

16 $-7 \times -6 =$ ______ [1]

17 $342 \div -9 =$ ______ [1]

18 $-6 \times -5 \times 12 =$ ______ [1]

19 $-17 + (-15) =$ ______ [1]

20 $-4 - 9 \times -2 =$ ______ [1]

21 $(13 + (-7)) \times -3 =$ ______ [1]

22 $24 \div -4 \times -2 =$ ______ [1]

23 $\frac{20 - 8}{6 - 8} =$ ______ [1]

24 $(-9 - 5) \div (-2 + 9) =$ ______ [1]

25 $6 - -18 \div 3 \times -2 + 4 \times -5 =$ ______ [1]

PART C Show all working.

26 Given that $6 \times 7 \times 8 \times 9 = 3024$, find the value of:

a $-6 \times 7 \times -8 \times -9$ — 1 mark

b -54×-56 — 1 mark

c $-3024 \div (-7 \times 8 \times -9)$ — 1 mark

d $3024 \div -42$ — 1 mark

27 The temperature in a freezer was –19 °C. It increased by 4 °C every hour for 6 hours.

a What was the temperature in the freezer at the end of the 6 hours? — 2 marks

b If the temperature then decreased by 3 °C every hour, how long would it take to return to its original temperature? — 1 mark

28 Find the value of:

a $-583 + 476 \times -2$ — 2 marks

b $\frac{-3 + 12 \times -2}{-2 - 3 - 4}$ — 2 marks

c $-16 - \frac{5 - 7}{-3 + 1} - 33$ — 2 marks

d $428 \div -4 \times 57 + 43 \times (-535 \div 5)$ — 2 marks

Total marks /40

LEVEL 2 TEST PART A

Time allowed: 40 minutes

Marks

1 The sum of –2 and –3 is:
Ⓐ –5 Ⓑ –6 Ⓒ 5 Ⓓ 6 [1]

2 The product of –3 and 4 is:
Ⓐ –7 Ⓑ –12 Ⓒ 1 Ⓓ 12 [1]

3 $3 \times (-4)^2 =$
Ⓐ –48 Ⓑ 48 Ⓒ –144 Ⓓ 144 [1]

4 $-3 + 7 \times -9 =$
Ⓐ –66 Ⓑ –36 Ⓒ 90 Ⓓ 36 [1]

5 $144 \div -6 =$
Ⓐ 150 Ⓑ 138 Ⓒ 24 Ⓓ –24 [1]

6 $5 - -20 \div -5 =$
Ⓐ –5 Ⓑ 5 Ⓒ 1 Ⓓ 9 [1]

7 $-37 + (-19) - 31 =$
Ⓐ –87 Ⓑ –49 Ⓒ –25 Ⓓ 13 [1]

8 $-56 \div 8 \times -7 =$
Ⓐ 1 Ⓑ –1 Ⓒ 49 Ⓓ –49 [1]

9 $4554 \div -3 \times (-2 - -5) =$
Ⓐ 414 Ⓑ 10 626 Ⓒ 4554 Ⓓ –4554 [1]

10 $-17 + 9 \times -5 - -23 =$
Ⓐ –85 Ⓑ –39 Ⓒ –36 Ⓓ 10 [1]

11 $63 \div -9 - 58 \div -2 =$
Ⓐ –36 Ⓑ –22 Ⓒ 36 Ⓓ 22 [1]

12 $-1800 \div -15 \times 6 =$
Ⓐ 20 Ⓑ 120 Ⓒ 320 Ⓓ 720 [1]

13 $(-23 + 17) \times (-36 - 8) =$
Ⓐ 208 Ⓑ 264 Ⓒ –264 Ⓓ –643 [1]

14 $\dfrac{-18 + 6 \times -5}{24 - 12 \times -3}$
Ⓐ 4 Ⓑ $\frac{1}{5}$ Ⓒ $-\frac{4}{5}$ Ⓓ $16\frac{3}{4}$ [1]

15 $-9 - 3 \times -7 + (-5) \times 17 - 42 =$
Ⓐ –43 Ⓑ 55 Ⓒ –115 Ⓓ –157 [1]

The Four Operations with Integers Calculator Allowed (Level 2 Test)

PART B Write answers only in the space provided.

16 $-48 - 16 \div -4 =$ ____________ [1]

17 $-28 - 15 \times 7 - 31 =$ ____________ [1]

18 $56 \div -4 \times -2 + 1 =$ ____________ [1]

19 $(9 - -16) \div (-3 - 2) =$ ____________ [1]

20 $13 \times (7 - 12) =$ ____________ [1]

21 $-7 + \dfrac{24}{2 - 6} =$ ____________ [1]

22 $-6 \times -6 \times -6 \times -5 \times -4 \times -8 \times -8 =$ ____________ [1]

23 $-19 \times 5 - 17 \times -6 =$ ____________ [1]

24 $\dfrac{-37 \times 9}{111 \div -3} =$ ____________ [1]

25 $-5 + (-6) \times -4 - 11 \times 8 - -9 \times 7 =$ ____________ [1]

PART C Show all working.

26 The table shows the balance of the account of a new business at the end of each of the first four months of trading.

Month	July	August	September	October
Balance	–$3160	–$1748	–$975	$648

a What does the negative value in some of the balances mean? [1 mark]

____

b What is the difference between the balance for August and September? [1 mark]

____

c What is the difference between the balance for July and October? [1 mark]

____

d What was the average balance for the four months? [2 marks]

____

____

e How much would the balance need to be in November to have a zero average balance for the first five months? [1 mark]

____

27 Hugo did this question $-7 + 9 \times 5 - -6 \times -2$ on his calculator.

a What should the answer have been? [1 mark]

____

b Hugo's calculator does not follow the correct order of operations and did the whole question in the order shown from left to right. What answer did Hugo get? [1 mark]

____

c Rewrite the question adding grouping symbols that would make the question correct using Hugo's calculator. [1 mark]

____

28 Insert grouping symbols to make these correct:

a $-48 \div 6 \times -2 = 4$ [1 mark]

b $7 - -5 \times 4 - 6 - 3 = -60$ [1 mark]

c $-3 \times -8 - 6 - -1 + 12 \div -4 = 20$ [1 mark]

29 a Find the value of $\dfrac{-6 \times (4 - 7) \div (-2 + 8)}{12 - 28 \div -7 + 5 \times -5}$. [1 mark]

____

b Rose did this question and had all the signs and operators correct but mistakenly changed both 5s to 4s. What answer did Rose get on her calculator? Briefly explain why. [2 marks]

____

____

Total marks /40

PRETEST

Time allowed: 15 minutes **Total marks: 15**

	Marks
1 What is 0.8 as a fraction in simplest form?	
(A) $\frac{1}{8}$ (B) $\frac{1}{3}$ (C) $\frac{17}{20}$ (D) $\frac{4}{5}$	1
2 What is the place value of 7 in 23.4795?	
(A) tens (B) tenths (C) hundredths (D) thousandths	1
3 0.3 + 0.07 =	
(A) 0.307 (B) 0.37 (C) 0.10 (D) 0.01	1
4 What is half of three-quarters?	
(A) $\frac{3}{2}$ (B) $\frac{3}{8}$ (C) $1\frac{1}{2}$ (D) $1\frac{1}{4}$	1
5 When rounded to three decimal places 12.378 64 is:	
(A) 12.4 (B) 12.378 (C) 12.379 (D) 12.864	1
6 As a decimal $\frac{1}{4}$ is:	
(A) 0.4 (B) 0.14 (C) 0.25 (D) 0.04	1
7 $\frac{5}{8}$ of 40 =	
(A) 10 (B) 15 (C) 20 (D) 25	1
8 $0.2 \times 0.3 =$	
(A) 0.05 (B) 0.06 (C) 0.5 (D) 0.6	1
9 $0.07 \div 100 =$	
(A) 0.0007 (B) 0.007 (C) 0.7 (D) 7	1
10 $1.3 - 0.48 =$	
(A) 0.82 (B) 0.88 (C) 0.92 (D) 1.18	1
11 $\frac{2}{5} + \frac{3}{7} =$	
(A) $\frac{5}{12}$ (B) $\frac{1}{2}$ (C) $\frac{31}{35}$ (D) $\frac{29}{35}$	1
12 $0.6 \div 0.3 =$	
(A) 2 (B) 0.2 (C) 0.02 (D) 0.002	1
13 $2.4 \times 1000 =$	
(A) 2004 (B) 2400 (C) 2.004 (D) 0.0024	1
14 $\frac{4}{5} \times \frac{2}{3} =$	
(A) $\frac{3}{4}$ (B) $\frac{2}{5}$ (C) $\frac{8}{15}$ (D) $\frac{14}{15}$	1
15 $8.0642 \div 4 =$	
(A) 2.165 (B) 2.1605 (C) 2.0165 (D) 2.016 05	1

Total marks ___ / 15

LEVEL 2 TEST PART A

Time allowed: 40 minutes

Marks

1 What is $\frac{1}{3}$ as a decimal?

(A) 0.13 (B) 0.03 (C) 0.3 (D) $0.\dot{3}$ — 1

2 7.32 + 0.8 =

(A) 7.4 (B) 7.04 (C) 8.12 (D) 7.112 — 1

3 Which is **not** a rational number?

(A) $\frac{3}{79}$ (B) 0.9 (C) $0.3\dot{1}$ (D) They are all rational. — 1

4 $\frac{1}{4} - \frac{1}{3} =$

(A) $-\frac{1}{12}$ (B) $-\frac{1}{7}$ (C) $\frac{1}{12}$ (D) $\frac{1}{7}$ — 1

5 $0.3\dot{8} =$

(A) 0.383 838 38… (B) 0.388 8888… (C) 0.838 383 83… (D) 0.388 388 388… — 1

6 $\frac{3}{4}$ of 0.2 =

(A) 0.15 (B) 0.275 (C) 0.015 (D) 0.075 — 1

7 $\frac{2}{9}$ as a decimal is:

(A) 0.29 (B) 0.222 222… (C) 0.45 (D) 0.292 929 29… — 1

8 $-1.2 \div -3 \times -2 =$

(A) –0.8 (B) 0.8 (C) –0.2 (D) 0.2 — 1

9 $0.0\dot{4}\dot{5} =$

(A) 0.045 045 045… (B) 0.045 555 55… (C) 0.045 454 545… (D) 0.044 444 445 — 1

10 –0.4 – 0.6 =

(A) 0.2 (B) –0.10 (C) –0.2 (D) –1 — 1

11 $-\frac{2}{3} \times \frac{4}{7} =$

(A) $\frac{2}{21}$ (B) $-\frac{2}{21}$ (C) $-\frac{6}{21}$ (D) $-\frac{8}{21}$ — 1

12 0.876 767 676 76… =

(A) $0.87\dot{6}$ (B) $0.8\dot{7}\dot{6}$ (C) $0.\dot{8}7\dot{6}$ (D) none of these — 1

13 $-0.2 \times 0.3 \times -0.4 =$

(A) –0.024 (B) –0.24 (C) 0.024 (D) 0.24 — 1

14 $0.9 - 0.6 \div 0.2 =$

(A) –2.1 (B) 0.15 (C) 1.5 (D) 2.1 — 1

15 $\frac{1}{2} + \frac{1}{3} \times \frac{1}{4} =$

(A) $\frac{1}{14}$ (B) $\frac{7}{12}$ (C) $\frac{5}{8}$ (D) $\frac{4}{7}$ — 1

Rational and Irrational Numbers Non-calculator **(Level 2 Test)**

PART B Write answers only in the space provided.

16 Is π rational or irrational? ______________ | 1

17 Write $0.3\dot{5}$ correct to four decimal places. ______________ | 1

18 When expressed as a decimal does $\frac{5}{8}$ terminate or repeat? ______________ | 1

19 What is $0.\dot{6}$ as a simple fraction? ______________ | 1

20 Is $\sqrt{9}$ rational or irrational? ______________ | 1

21 $-0.9 \div 0.03 =$ ______________ | 1

22 Is $0.2\dot{0}9\dot{7}$ rational or irrational? ______________ | 1

23 What is $\frac{1}{6}$ as a decimal? ______________ | 1

24 Is $\sqrt{5}$ rational or irrational? ______________ | 1

25 $\frac{3}{5} - 1\frac{2}{5} =$ ______________ | 1

PART C Show all working.

26 a Write $0.0\dot{9}\dot{6}$ showing how its digits repeat. [1 mark]

____

b Write $0.0\dot{9}\dot{6}$ correct to four decimal places. [1 mark]

____

c What is the answer to part **a** as an improper fraction? [1 mark]

____

d What is $\frac{-2.8 + 5.4}{13}$ as both a fraction and decimal? [2 marks]

____

____

27 a Find $1\frac{2}{3} \times 2\frac{1}{4}$ giving the answer as a mixed number in simplest form. [2 marks]

____

____

b What is the answer to part **a** as a decimal? [1 mark]

____

c Briefly explain why the answer is rational. [1 mark]

____

28 a What is $-2.8 + 5.4$? [1 mark]

____

b What is the answer to part **a** as a mixed number in simplest form? [1 mark]

____

29 Craig's calculator gives the value 1.732 050 808 for $\sqrt{3}$. Craig wrote the answer as $1.732\ 05\dot{0}\dot{8}$. Is that correct? Explain why or why not. [2 marks]

____

____

____

30 Express $\frac{7}{11}$ as a repeating decimal. [2 marks]

____

____

Total marks /40

LEVEL 2 TEST PART A

Time allowed: 40 minutes

Marks

1 Which is a rational number? [1]

Ⓐ $\sqrt{2}$ Ⓑ π Ⓒ $\sqrt{5}$ Ⓓ $\sqrt{9}$

2 When expressed as a decimal, which of these is a recurring decimal? [1]

Ⓐ $\frac{3}{4}$ Ⓑ $\frac{3}{5}$ Ⓒ $\frac{3}{7}$ Ⓓ $\frac{3}{8}$

3 $0.\dot{1}5\dot{7} =$ [1]

Ⓐ 0.157 157 157… Ⓑ 0.157 171 717… Ⓒ 0.157 575 757… Ⓓ 0.157 777 77…

4 $5.7 - -2.84 + 6.21 \times -3.7 =$ [1]

Ⓐ –14.437 Ⓑ –20.117 Ⓒ 31.517 Ⓓ –54.575

5 As a decimal, $\frac{1}{18} =$ [1]

Ⓐ 0.05 Ⓑ $0.0\dot{5}$ Ⓒ $0.\dot{0}\dot{5}$ Ⓓ 0.055 5556

6 What is the value of π correct to three decimal places? [1]

Ⓐ 0.314 Ⓑ 3.14 Ⓒ 3.141 Ⓓ 3.142

7 $0.\dot{1}\dot{2} =$ [1]

Ⓐ $\frac{2}{165}$ Ⓑ $\frac{3}{25}$ Ⓒ $\frac{4}{33}$ Ⓓ $\frac{11}{90}$

8 $1\frac{3}{4} + 2\frac{5}{6} \times 3\frac{1}{2} =$ [1]

Ⓐ $16\frac{1}{24}$ Ⓑ $12\frac{7}{24}$ Ⓒ $11\frac{2}{3}$ Ⓓ $8\frac{1}{6}$

9 What is $\frac{7}{24}$ as a decimal to three decimal places? [1]

Ⓐ 0.292 Ⓑ 0.296 Ⓒ 0.297 Ⓓ 0.724

10 Which of these is irrational? [1]

Ⓐ $\sqrt{1521}$ Ⓑ $\sqrt{2209}$ Ⓒ $\sqrt{2306}$ Ⓓ $\sqrt{2704}$

11 0.235 8585 858 58… = [1]

Ⓐ $0.235\dot{8}$ Ⓑ $0.23\dot{5}\dot{8}$ Ⓒ $0.2\dot{3}\dot{5}\dot{8}$ Ⓓ $0.\dot{2}35\dot{8}$

12 $0.6\dot{8}$ correct to three decimal places is: [1]

Ⓐ 0.686 Ⓑ 0.687 Ⓒ 0.688 Ⓓ 0.689

13 Which is **not** a rational number? [1]

Ⓐ fraction Ⓑ repeating decimal Ⓓ terminating decimal Ⓓ They are all rational.

14 $\frac{17}{36} =$ [1]

Ⓐ 0.472 Ⓑ $0.47\dot{2}$ Ⓒ $0.4\dot{7}\dot{2}$ Ⓓ $0.\dot{4}\dot{7}\dot{2}$

15 $\frac{0.16 \div -0.04}{-0.2 \times 0.08} =$ [1]

Ⓐ 250 Ⓑ 1.6 Ⓒ 0.25 Ⓓ 0.064

PART B Write answers only in the space provided.

16 What is $0.\dot{3}\dot{6}$ as a fraction in simplest form? ______________ 1

17 What is $\frac{5}{8}$ as a decimal? ______________ 1

18 When expressed as a decimal, does $\frac{16}{27}$ repeat or terminate? ______________ 1

19 What is $\sqrt{13}$ to five decimal places? ______________ 1

20 Is $\sqrt{13}$ rational or irrational? ______________ 1

21 $1.8 - 3.6 \div -0.04 =$ ______________ 1

22 Write $0.0\dot{4}\dot{7}$ correct to six decimal places. ______________ 1

23 $-3\frac{1}{4} - 2.4$ as a decimal = ______________ 1

24 $0.\dot{6} + 0.\dot{7} =$ ______________ 1

25 $-2\frac{1}{3} \div \frac{3}{5} \times 1\frac{7}{8}$ as a mixed numeral in simplest form = ______________ 1

PART C Show all working.

26 **a** What is two-thirds of $\frac{29}{37}$ as a single fraction? 1 mark

____

b What is the answer to part **a** expressed as a recurring decimal? 1 mark

____

c What is the answer to part **a** correct to six decimal places? 1 mark

____

27 **a** Find $\frac{6.4 - 3.2 \times 1.8}{6.1 - 15.7}$, giving the answer as a decimal. 1 mark

____

b Give the answer to part **a** as a fraction in simplest form. 1 mark

____

28 A calculator shows that $\frac{5}{17} = 0.294\ 117\ 6471$.

a Is $\frac{5}{17}$ rational? Justify your answer. 2 marks

____

____

b The tenth digit after the decimal point shown on the calculator is not correct in the actual value of $\frac{5}{17}$. In that case what should the tenth digit be and why does the calculator show it as a 1? 2 marks

____

____

c In actual fact $\frac{5}{17}$ has the maximum possible number of repeating digits. How many repeating digits must it have and why must that be the maximum possible number? 2 marks

____

____

____

29 **a** Find $\frac{5}{8} - 1\frac{3}{5} \times 3\frac{2}{3} - \frac{1}{9}$ giving the answer as a mixed numeral in simplest form. 1 mark

____

b Give the answer to part **a** as a decimal. 1 mark

____

c Give the answer rounded to four decimal places. 1 mark

____

d What is the difference between the answers to part **b** and part **c**? Give the answer as a decimal. 1 mark

____

Total marks /40

Indices Calculator Allowed

PRETEST

Time allowed: 15 minutes **Total marks: 16**

					Marks
1 $5^2 =$					
	(A) 52	(B) 32	(C) 25	(D) 10	1
2 $3^4 =$					
	(A) $3 + 3 + 3 + 3$	(B) $3 \times 3 \times 3 \times 3$	(C) $4 + 4 + 4$	(D) $4 \times 4 \times 4$	1
3 $6^3 =$					
	(A) 18	(B) 216	(C) 729	(D) none of these	1
4 In the expression 6^9, 6 is the:					
	(A) base	(B) index	(C) exponent	(D) power	1
5 $7 \times 7 \times 7 \times 7 \times 7 =$					
	(A) 5^7	(B) 7^5	(C) 35	(D) none of these	1
6 $3^2 + 4^2 =$					
	(A) 2401	(B) 49	(C) 25	(D) none of these	1
7 $(-4)^2 =$					
	(A) -16	(B) -8	(C) 8	(D) 16	1
8 $(-2)^3 =$					
	(A) -8	(B) -6	(C) 6	(D) 8	1
9 $2 + 3^2 =$					
	(A) 10	(B) 11	(C) 25	(D) 32	1
10 $5^3 =$					
	(A) $5 \times 3 \times 3$	(B) $5 \times 5 \times 3$	(C) $5 \times 5 \times 5$	(D) $3 \times 3 \times 3 \times 3 \times 3$	1
11 $10 - 2^3 =$					
	(A) 512	(B) 6	(C) 4	(D) 2	1
12 $2^3 \times 3^2 =$					
	(A) $2 \times 2 \times 2 \times 3 \times 3$	(B) $2 \times 3 \times 3 \times 2$	(C) $(2 + 2 + 2) \times (3 + 3)$	(D) $2 \times 2 \times 2 + 3 \times 3$	1
13 $2 \times 7^2 =$					
	(A) 392	(B) 196	(C) 98	(D) 28	1
14 $10^2 - 6^2 =$					
	(A) 4^2	(B) 5^2	(C) 6^2	(D) 8^2	1
15 $2^3 + 4 \times 3^2 =$					
	(A) 1296	(B) 152	(C) 108	(D) 44	1
16 $3^4 - 5 \times 2^2 =$					
	(A) 61	(B) 304	(C) 5041	(D) -17	1

Total marks / 16

LEVEL 1 TEST PART A

Time allowed: 40 minutes

					Marks
1	$5^3 \times 5^4 =$				
	(A) 5^7	(B) 5^{12}	(C) 25^7	(D) 25^{12}	1
2	$7^9 \div 7^3 =$				
	(A) 1^3	(B) 1^6	(C) 7^3	(D) 7^6	1
3	$(3^2)^4 =$				
	(A) 3^6	(B) 3^8	(C) 9^6	(D) 9^8	1
4	$5^0 =$				
	(A) 0	(B) 1	(C) 5	(D) none of these	1
5	$12^6 \div 12 =$				
	(A) 1^6	(B) 12^3	(C) 12^5	(D) 12^6	1
6	$6^3 \times 6^3 =$				
	(A) 12^3	(B) 2×6^3	(C) 6^6	(D) 6^9	1
7	$3^4 + 3^5 =$				
	(A) 3^9	(B) 6^9	(C) 9^{20}	(D) none of these	1
8	$2^0 + 3^0 + 4^0 =$				
	(A) 9	(B) 3	(C) 1	(D) 0	1
9	$2^2 \times 3^2 =$				
	(A) 5^4	(B) 5^2	(C) 6^4	(D) 6^2	1
10	$3^5 \times 3^3 \times 3^2 =$				
	(A) 3^{10}	(B) 3^{30}	(C) 27^{10}	(D) 27^{30}	1
11	$\frac{10^9}{10^3} =$				
	(A) 1^6	(B) 10^3	(C) 10^6	(D) 100^3	1
12	$(5^3)^3 =$				
	(A) 25^3	(B) 5^6	(C) 5^9	(D) 5^{27}	1
13	$2^3 + 2^2 =$				
	(A) 2^5	(B) 2^6	(C) 4^5	(D) none of these	1
14	$3 \times 3^4 =$				
	(A) 3^4	(B) 3^5	(C) 3^6	(D) 3^8	1
15	$2^5 \times 3^5 =$				
	(A) 6^5	(B) 6^{10}	(C) 5^5	(D) 5^{10}	1

Indices Non-calculator (Level 1 Test)

PART B Write answers only in the space provided.

16 $2^9 \times 2^6$ in index form = ________________ [1]

17 $3^{15} \div 3^5$ in index form = ________________ [1]

18 $(2^5)^4$ in index form = ________________ [1]

19 6^0 = ________________ [1]

20 What is the value of $(-2)^3$? ________________ [1]

21 $5^3 \times 5^2 \times 5^4$ in index form = ________________ [1]

22 $10^9 \div 10^3 \div 10$ in index form = ________________ [1]

23 $3 \times (3^4)^3$ in index form = ________________ [1]

24 What is the value of 5×2^0? ________________ [1]

25 $7^8 \div 7^2 \times 7^4$ in index form = ________________ [1]

PART C Show all working.

26 Find, giving answers in simplest index form:

a $\dfrac{3^8 \times 3^4 \times 3^3}{3^6 \times 3^5 \times 3^2}$ [2 marks]

____

____

b $(5^4 \times 5^7)^3 \div (5^6 \div 5)$ [2 marks]

____

____

c $(2^7)^3 \times (2^5)^4$ [2 marks]

____

____

27 What is the value of:

$3 \times 5^0 + (3 \times 5)^0 + 3^0 \times 5^0$? [2 marks]

____

____

28 Write in simplest index form:

a 9^2 [1 mark]

____

b 25^3 [1 mark]

____

c 16^3 [2 marks]

____

29 a What is the value of 2^3? [1 mark]

____

b Using the answer to part **a**, what is the value of $2^3 \div 2^3$? [1 mark]

____

c Explain why using the index laws with $2^3 \div 2^3$ leads to a value for 2^0. [1 mark]

____

____

Total marks /40

LEVEL 2 TEST PART A

Time allowed: 40 minutes

					Marks
1	$12^7 \times 12^8 =$				
	(A) 12^{15}	(B) 12^{56}	(C) 144^{15}	(D) 144^{56}	1
2	$10^8 \div 10^2 =$				
	(A) 10^6	(B) 10^4	(C) 1^4	(D) none of these	1
3	$2 \times 3^0 =$				
	(A) 0	(B) 1	(C) 2	(D) 6	1
4	$2^6 \times 2^5 \times 2 =$				
	(A) 2^{11}	(B) 2^{12}	(C) 2^{31}	(D) 8^{11}	1
5	$3^0 + 4^0 + 5^0 =$				
	(A) 0	(B) 1	(C) 3	(D) 12	1
6	$3^8 \div 3^4 =$				
	(A) 81	(B) 12	(C) 9	(D) none of these	1
7	$5^4 + 5^3 =$				
	(A) 5^7	(B) 5^{12}	(C) 10^7	(D) none of these	1
8	$(2^5)^2 \times 2 =$				
	(A) 2^{11}	(B) 2^8	(C) 4^{11}	(D) 4^8	1
9	$(5^0)^2 =$				
	(A) 0	(B) 1	(C) 5	(D) 25	1
10	$((-5)^2)^3 =$				
	(A) $-(5^5)$	(B) $-(5^6)$	(C) 5^5	(D) 5^6	1
11	$4^2 \times 2 =$				
	(A) 2^3	(B) 2^4	(C) 2^5	(D) 2^6	1
12	$2^0 \times 3^0 =$				
	(A) 0	(B) 1	(C) 2	(D) 6	1
13	$(2^2 \times 2^5)^3 =$				
	(A) 2^{10}	(B) 2^{13}	(C) 2^{21}	(D) 2^{30}	1
14	$(7^3)^0 =$				
	(A) 0	(B) 1	(C) 7	(D) 7^3	1
15	$3^2 \times 2^5 =$				
	(A) 5^7	(B) 6^7	(C) 6^{10}	(D) none of these	1

Indices Non-calculator **(Level 2 Test)**

PART B Write answers only in the space provided.

16 What is 9^5 in simplest index form? ________________ 1

17 Evaluate $3^0 + 4 \times 2^3$. ________________ 1

18 What is the value of $(2^0)^4$? ________________ 1

Answer questions 19 to 25 in index form.

19 $11^3 \times 11^2 \times 11 =$ ________________ 1

20 $(3^7)^5 =$ ________________ 1

21 $7^6 \times 7^3 \times 7^2 =$ ________________ 1

22 $\dfrac{6^8}{6^2 \times 6^3} =$ ________________ 1

23 $5^{12} \div 5^3 \div 5^2 =$ ________________ 1

24 $10^4 \times 10^6 \div 10^3 =$ ________________ 1

25 $2^{10} \div 2^2 \times 2^4 =$ ________________ 1

PART C Show all working.

26 Find in simplest index form:

a $(2^8 \div 2^4)^2 \times (2^{10} \div 2^2)^3$ 2 marks

____

____

b $(3^4)^3 \times (3^5)^2 \div (3^2)^6$ 2 marks

____

____

c $4^3 \times 8^2$ 2 marks

____

____

d $\dfrac{5^3 \times 7^5 \times 5^8 \times 7}{7^9 \times 5^4 \div 7^3}$ 2 marks

____

____

e $5^2 + 5^2 + 5^2 + 5^2 + 5^2$ 2 marks

____

____

27 Find the value of:

a $7^2 - 7 \times 7^0$ 1 mark

____

b $-3^2 + (5^2)^0 + (2^2)^2$ 2 marks

____

____

c $(10^{12} \div 10^4 \times 10^2) \div 10^3$ 2 marks

____

____

Total marks /40

LEVEL 3 TEST PART A

Time allowed: 40 minutes

Marks

1 $4^5 =$

(A) 2^7 (B) 2^8 (C) 2^9 (D) 2^{10} [1]

2 $5^7 \times 5^0 =$

(A) 0 (B) 1 (C) 5^7 (D) 5^8 [1]

3 $8^3 \div 2^3 =$

(A) 2^3 (B) 2^6 (C) 4^0 (D) 4^1 [1]

4 $3 + 3^4 \times 3^4 =$

(A) 3^9 (B) 18^4 (C) 3^{17} (D) none of these [1]

5 $7^0 - 2^0 =$

(A) 0 (B) 1 (C) 5 (D) none of these [1]

6 $(-2)^2 \times (-2)^3 =$

(A) $(-2)^5$ (B) 2^5 (C) $(-2)^6$ (D) 2^{10} [1]

7 $5^8 \div 5^2 \times 5^3 =$

(A) 5^3 (B) 5^7 (C) 5^9 (D) 5^{12} [1]

8 $9^6 \div 3^2 =$

(A) 3^4 (B) 3^6 (C) 3^8 (D) 3^{10} [1]

9 $\frac{3^8 \times 2^3}{3^5} =$

(A) 6^3 (B) 6^6 (C) 6^9 (D) 6^{19} [1]

10 $2^7 + 2^7 =$

(A) 4^7 (B) 2^{14} (C) 2^8 (D) none of these [1]

11 $(5^4 \times 5^6)^0 =$

(A) 0 (B) 1 (C) 5^{10} (D) 5^{100} [1]

12 $(3^2)^3 \times 3^4 =$

(A) 3^9 (B) 3^{10} (C) 3^{20} (D) 3^{24} [1]

13 $(6^8 \div 6^2)^4$

(A) 6^{24} (B) 6^{16} (C) 6^{10} (D) 6^8 [1]

14 $2^6 \times 4^3 =$

(A) 2^9 (B) 2^{11} (C) 8^4. (D) 8^9 [1]

15 $9^{12} \div 9^4 =$

(A) 3^3 (B) 3^6 (C) 3^8 (D) 3^{16} [1]

Indices Non-calculator **(Level 3 Test)**

PART B Write answers only in the space provided.

16 What is the value of $7 \times 6^0 + 6 \times 7^0$? ________________ 1

17 $(2^3 \times 2^5)^2$ in index form = ________________ 1

18 What is 4^7 in simplest index form? ________________ 1

19 Evaluate $(3^0)^0$. ________________ 1

20 Find the value of $7^2 \times 7^5 \div 7^6$. ________________ 1

21 Evaluate $(-3)^8 \div (-3)^5$. ________________ 1

22 $\dfrac{6^2 \times 6^7}{6^9 \div 6^3}$ in index form = ________________ 1

23 Evaluate $6^5 \div 3^5$. ________________ 1

24 $(5^3)^2 \times (5^2)^4$ in index form = ________________ 1

25 $17^{10} \div 17^6 \times 17^8 \div 17$ in index form = ________________ 1

PART C Show all working.

26 Write in simplest index form:

a $\dfrac{27^6}{9^5}$ 2 marks

____

____

b $2^7 \times 2^8 + 2^6 \times 2^9$ 2 marks

____

____

c $\dfrac{\left((7^4)^2 \times 7^6\right)^3}{(7^{15} \div 7^3)^2}$ 2 marks

____

____

27 Evaluate:

a $(3^8 \times 3^6 \div 3^0 \div 3^{12})^2$ 1 mark

____

b $(5^5 \div 5^2)^2 \times (5^3 \times 5^3)^3 \div (5^4)^6$ 2 marks

____

____

c $2^8 \times 3^2 \times 5^5$ 2 marks

____

____

28 a Express 108 as a product of its prime factors in index form. 1 mark

____

b Now express 108^4 as the product of its prime factors in index form. 1 mark

____

29 Nancy said that $3^5 \times 3^3 = 9^{15}$. Explain why she is wrong, give the correct answer and show why that is the correct answer. 2 marks

____

____

Total marks /40

Percentages

PRETEST

Time allowed: 15 minutes **Total marks: 15**

					Marks
1	What percentage is $\frac{1}{4}$?				
	(A) 14%	(B) 20%	(C) 25%	(D) 40%	1
2	What is 10% of $12.60?				
	(A) $1.26	(B) $12.06	(C) $12.60	(D) $11.34	1
3	What is 70% as a decimal?				
	(A) 70.0	(B) 0.7	(C) 0.07	(D) 0.007	1
4	What percentage is 60 out of 200?				
	(A) 60%	(B) 120%	(C) 12%	(D) 30%	1
5	What is 80% as a simple fraction?				
	(A) $\frac{1}{80}$	(B) $\frac{1}{8}$	(C) $\frac{4}{50}$	(D) $\frac{4}{5}$	1
6	What percentage is 45 minutes of an hour?				
	(A) 45%	(B) 60%	(C) 75%	(D) 85%	1
7	What is 0.04 as a percentage?				
	(A) 4%	(B) 14%	(C) 25%	(D) 40%	1
8	30% of 50 =				
	(A) 6	(B) 8	(C) 15	(D) 20	1
9	What percentage is 56 out of 80?				
	(A) 76%	(B) 70%	(C) 64%	(D) 56%	1
10	$12\frac{1}{2}\%$ =				
	(A) 0.125	(B) 0.0125	(C) 1.25	(D) 0.1205	1
11	What is 30% as a simple fraction?				
	(A) $\frac{1}{30}$	(B) $\frac{1}{3}$	(C) $\frac{3}{100}$	(D) $\frac{3}{10}$	1
12	What is 70% of $50 000?				
	(A) $3500	(B) $35 000	(C) $1400	(D) $14 000	1
13	What percentage is 150 out of 100?				
	(A) 150%	(B) 15%	(C) 60%	(D) $66\frac{2}{3}\%$	1
14	What is $\frac{3}{5}$ as a percentage?				
	(A) 35%	(B) 53%	(C) 60%	(D) 70%	1
15	As a decimal, 350% =				
	(A) 0.35	(B) 0.035	(C) 350.0	(D) 3.5	1

Total marks /15

LEVEL 2 TEST PART A

Time allowed: 40 minutes

Marks

1 What is 50% of $128?

(A) $78 (B) $56 (C) $25.60 (D) $64 [1]

2 If a painting is bought for $1200 and sold for $2300, how much is the profit?

(A) $900 (B) $1100 (C) $1200 (D) $2300 [1]

3 What is the whole amount if 20% is $1600?

(A) $8000 (B) $7200 (C) $3200 (D) $800 [1]

4 If 1200 is increased by 40% the result is:

(A) 1240 (B) 1600 (C) 1680 (D) 3000 [1]

5 How much is a discount of 30% on an item priced at $15.00?

(A) 45c (B) $4.50 (C) $5.00 (D) 50c [1]

6 If $90 000 is decreased by 20% the result is:

(A) $18 000 (B) $88 200 (C) $70 000 (D) $72 000 [1]

7 What is the percentage increase if $50 is increased to $80?

(A) 60% (B) 62.5% (C) 30% (D) 37.5% [1]

8 What is 250% of $800?

(A) $200 (B) $2000 (C) $2800 (D) $3200 [1]

9 Multiplying by which of these will decrease an amount by 5%?

(A) 0.05 (B) 0.5 (C) 0.8 (D) 0.95 [1]

10 What is the percentage increase if $40 is increased by $20?

(A) 50% (B) $66\frac{2}{3}\%$ (C) $33\frac{1}{3}\%$ (D) 200% [1]

11 What price will you pay for a shirt marked at $70.00 if you receive a 20% discount?

(A) $35.00 (B) $65.00 (C) $50.00 (D) $56.00 [1]

12 If an item was bought for $20 000 and sold for $25 000, what was the profit as a percentage of the cost price?

(A) 5% (B) 20% (C) 25% (D) 30% [1]

13 What is the result if $600 is increased by 150%?

(A) $900 (B) $1200 (C) $1500 (D) $1600 [1]

14 What is the percentage decrease if an item marked at $80 is sold for $60?

(A) 5% (B) 20% (C) 25% (D) 40% [1]

15 A car was bought for $20 000 and sold at a loss of 15%. What was the selling price?

(A) $18 000 (B) $17 000 (C) $16 000 (D) $15 000 [1]

PART B Write answers only in the space provided.

16 What is 70% of $400? ________ [1]

17 What percentage is 120 out of 300? ________ [1]

18 What is the whole amount if 5% is $40? ________ [1]

19 Increase $120 by 20%. ________ [1]

20 Decrease $3000 by 10%. ________ [1]

21 Increase $80 by 200%. ________ [1]

22 A price increases from $50 to $60. What is the percentage increase? ________ [1]

23 A watch bought for $2000 was sold for a profit of 8%. What was the selling price? ________ [1]

24 A person's weight decreased from 80 kg to 72 kg. What percentage decrease is that? ________ [1]

25 Multiplying by 1.1 will increase an amount by what percentage? ________ [1]

PART C Show all working.

26 Brian bought a bike for $1200 and sold it for $900.

a What was the loss? [1 mark]

____

b What was the loss as a percentage of the cost price? [1 mark]

____

c Brian made a profit of 20% on the cost price of another bike that he sold at the same time. If he sold the second bike for $1200, what did it cost him? [1 mark]

____

27 A house bought for $200 000 was later sold for half a million dollars.

a What was the profit? [1 mark]

____

b What was the percentage profit on the cost price? [1 mark]

____

28 a In a netball game, the goal shooter scored 34 goals from 40 attempts. What percentage of the attempts was successful? [1 mark]

____

b In the next game the goal shooter had a 90% success rate. If she scored 27 goals, how many attempts at goal were made? [1 mark]

____

29 Some friends contributed different amounts of money to a project that they worked on together. Jason put in 20%, Meg put in 25%, Eli contributed 40% and Zane contributed the rest.

a What percentage did Zane contribute? [1 mark]

____

b If Eli put in $1000, how much did Meg contribute? [2 marks]

____

____

c The project made a total of $3000. If Jason received 20% of the profit, how much did he get? [2 marks]

____

____

30 a The price of a washing machine increased by 5%. If the original price was $500, what was the new price? [1 mark]

____

b One store had a sale and sold the washing machine for $483. What was the percentage saving on the new price? [2 marks]

____

____

Total marks /40

Percentages Calculator Allowed

LEVELS 1 AND 2 TEST — PART A

Time allowed: 40 minutes

	Marks
1 What is 35% of $24 000? (A) $685 (B) $6850 (C) $8400 (D) $14 580	1
2 What percentage is 78 out of 120? (A) 59% (B) 61% (C) 63% (D) 65%	1
3 If 25% is 600 kg, what is 100%? (A) 150 kg (B) 4200 kg (C) 1500 kg (D) 2400 kg	1
4 If a tool is sold for $700 after being bought for $900, this is a: (A) loss of $200 (B) profit of $200 (C) loss of $700 (D) profit of $700	1
5 If $620 is increased by 30%, what is the new amount? (A) $920 (B) $806 (C) $638.60 (D) $650	1
6 If $500 increases to $1000 this is an increase of: (A) 50% (B) 75% (C) 100% (D) 200%	1
7 What is the whole amount if 80% is $180? (A) $144 (B) $200 (C) $220 (D) $225	1
8 If $1400 is decreased by 20% the new amount is: (A) $1120 (B) $1200 (C) $1372 (D) $1380	1
9 $2\frac{1}{4}$% of $10 000 is: (A) $250 (B) $225 (C) $2500 (D) $2250	1
10 Multiplying an amount by 1.5 will increase it by: (A) 5% (B) 15% (C) 50% (D) 150%	1
11 If a television is for sale for $550.00 including GST (10%), how much is the GST? (A) $5.00 (B) $5.50 (C) $50.00 (D) $55.00	1
12 What percentage is 125 of 75? (A) $166\frac{2}{3}$% (B) $66\frac{2}{3}$% (C) 60% (D) none of these	1
13 What will you pay for a computer priced at $480 if you receive a 5% discount? (A) $384 (B) $456 (C) $470 (D) $475	1
14 If a price is reduced from $580 to $500 the percentage reduction is closest to: (A) 14% (B) 16% (C) 18% (D) 20%	1
15 If 30% of an amount is $750, what is 80% of the amount? (A) $1800 (B) $2000 (C) $2100 (D) $2250	1

PART B Write answers only in the space provided.

16 72% of $19 = ________________ [1]

17 What percentage is 58 out of 80? ________________ [1]

18 If 60% is $360, what is the whole amount? ________________ [1]

19 Increase 20 m by 20%. ________________ [1]

20 Decrease $900 by 9%. ________________ [1]

21 An increase of 5% on $1800 is what final amount? ________________ [1]

22 A box is bought for $50 and sold for $60.
What is the profit as a percentage of the cost price? ________________ [1]

23 A car is bought for $30 000 and sold at a loss of 20%. What is the selling price? ________________ [1]

24 Increase $1200 by 300%. ________________ [1]

25 What is the percentage decrease if 8000 is reduced to 6000? ________________ [1]

PART C Show all working.

26 A boat was bought for $17 000 and sold for $15 000. What was the loss as a percentage of the cost price? Give the answer to the nearest 0.1%. [2 marks]

____

____

27 The price, including GST (10%), of a car was $32 560.

a How much was the GST? [1 mark]

____

b The buyer was allowed 25% off the price because she had traded in her old car. How much remained to be paid? [1 mark]

____

28 There are 147 students in Year 8 at a school. Of the school's students, 17.5% are in Year 8 and 15% are in Year 9. How many more students are in Year 8 than in Year 9? [2 marks]

____

____

29 A shop is increasing all its prices by 20%.

a By how much will the price of a watch previously priced at $720 increase? [1 mark]

____

b What will be the new price of a ring previously selling for $1680? [1 mark]

____

c The new price of a necklace is $2520. What was the old price? [2 marks]

____

____

d The shop buys some earrings for $75 and sells them for $97.50. What percentage profit does the shop make on the earrings? [2 marks]

____

____

e A bracelet was bought by the shop for $400. Under the old price it would have made a profit of 10%. What percentage profit will the shop make on the new price? [3 marks]

____

____

____

Total marks /40

Percentages Calculator Allowed

LEVELS 2 AND 3 TEST — PART A

Time allowed: 40 minutes

Question	Marks
1 What is $12\frac{1}{2}\%$ of \$9500? (A) \$1187.50 (B) \$1144.75 (C) \$760 (D) \$7600	1
2 What percentage is 81 of 60? (A) 65% (B) 74% (C) 126% (D) 135%	1
3 What is 32% of a distance if 72% is 1152 km? (A) 460 km (B) 512 km (C) 640 km (D) 832 km	1
4 If 900 g is increased by 13% the resulting amount is: (A) 783 g (B) 1017 g (C) 1200 g (D) 1683 g	1
5 13 050 000 is 87% of what amount? (A) 11 353 500 (B) 15 000 000 (C) 15 517 200 (D) 17 000 000	1
6 About what percentage is 15 out of 26? (A) 4% (B) 39% (C) 58% (D) 73%	1
7 What is 350% of \$1410? (A) \$493.50 (B) \$1903.50 (C) \$4935 (D) \$6345	1
8 If \$1700 is decreased by 2% the resulting amount is: (A) \$1360 (B) \$1565 (C) \$1615 (D) \$1666	1
9 An item is bought for \$3500 and sold for \$3920. This is a profit of what per cent? (A) 12% (B) 15% (C) 18% (D) 20%	1
10 If 312 increases to 624 what is the percentage increase? (A) 50% (B) 100% (C) 150% (D) 200%	1
11 What is the whole amount if 6.5% is 806? (A) 12 400 (B) 5239 (C) 52 390 (D) 13 322	1
12 A painting is sold at a profit of 20% on the cost price. If it sold for \$960, what was the cost price? (A) \$192 (B) \$768 (C) \$800 (D) \$912	1
13 If 450 is reduced to 420, what is the percentage decrease? (A) $6\frac{2}{3}\%$ (B) $7\frac{1}{7}\%$ (C) $3\frac{1}{3}\%$ (D) $9\frac{1}{3}\%$	1
14 A packet of soup states that each serve has 722 mg of sodium which is 31% of the recommended daily intake. The recommended daily intake of sodium is about: (A) 224 mg (B) 2170 mg (C) 2240 mg (D) 2330 mg	1
15 Martin saved \$250 when he bought a drone for \$1650. Which is closest to the percentage discount Martin received? (A) 15% (B) 14% (C) 13% (D) 12%	1

Percentages Calculator Allowed (Levels 2 and 3 Test)

PART B Write answers only in the space provided.

16 Find the percentage increase if \$72 is increased to \$96. ______________ [1]

17 What is 225% of \$72 000? ______________ [1]

18 What percentage is 1260 of 2160? ______________ [1]

19 What is $10\frac{1}{4}$% of 600 ha? ______________ [1]

20 What is the result if 150 is increased by 40%? ______________ [1]

21 Multiplying an amount by 1.075 will increase it by what percentage? ______________ [1]

22 What is the result if 90 is decreased by $2\frac{1}{2}$%? ______________ [1]

23 What is the whole amount if 19% is \$14.25? ______________ [1]

24 Find the percentage increase if \$12 is increased by \$16. ______________ [1]

25 What is the result if \$64 is increased by 175%? ______________ [1]

PART C Show all working.

26 a In a shop, a chair has been discounted by \$54 and sells for \$396. What percentage discount is this? [1 mark]

____

b If a table is discounted by the same percentage, what will you pay if it was originally priced at \$3200? [1 mark]

____

c The price of tableware at the shop has increased from \$30 to \$72. What percentage increase is that? [1 mark]

____

d The price of cushions, originally \$5 each, has been increased by 160%. What is the cost of five cushions at the increased price? [1 mark]

____

27 a In a certain town, 37 188 of the total population of 77 475 people are male. What percentage of the population is female? [1 mark]

____

b If 3.5% of the total population is 85 or over, how many people is that? (Give the answer to the nearest hundred.) [1 mark]

____

c If the population increased by 2.5% and the percentage of females remained the same, what would the new population be and about how many would be male? [2 marks]

____

____

28 There are four main blood groups. Around 38% of the population has blood group A, 10% group B, 49% group O and the rest have blood group AB.

a Given that the population of Australia is about 24 million, about how many people will have blood group AB? [2 marks]

____

____

b In each blood group, the blood is also said to be either positive ($^+$) or negative ($^-$). Around 2.16 million Australians have O^- blood. About what percentage of blood group O is O^-? [2 marks]

____

____

29 Ed bought a rare book for \$2400 and sold it to Annie, making a profit of 15%.

a How much did Annie pay for the book? [1 mark]

____

b Later Annie sold the book, making a loss of 15%. When told this, Ed thought that Annie must have sold the book for \$2400. Explain why Ed is wrong. [1 mark]

____

c For how much more or less than \$2400 did Annie sell the book? [1 mark]

____

Total marks /40

PRETEST

Time allowed: 15 minutes **Total marks: 15**

Marks

1 In simplest form, the ratio 8 : 12 is:

(A) 1 : 4 (B) 2 : 3 (C) 3 : 4 (D) 4 : 6 [1]

2 There are 15 men in a group of 29 people. What is the ratio of men to women?

(A) 15 : 29 (B) 29 : 15 (C) 14 : 15 (D) 15 : 14 [1]

3 How long will it take to travel 150 km at 60 km/h?

(A) $1\frac{1}{2}$h (B) $2\frac{1}{4}$h (C) $2\frac{1}{2}$h (D) $2\frac{3}{4}$h [1]

4 The ratio of blue to green tickets in a hat is 3 : 5. What fraction of the tickets is blue?

(A) $\frac{3}{8}$ (B) $\frac{5}{8}$ (C) $\frac{2}{5}$ (D) $\frac{3}{5}$ [1]

5 Which will **not** simplify to 2 : 5?

(A) 6 : 15 (B) 8 : 11 (C) 14 : 35 (D) 18 : 45 [1]

6 How far will you travel in 45 minutes at 80 km/h?

(A) 36 km (B) 50 km (C) 60 km (D) 75 km [1]

7 If the ratio of cars to trucks is 3 : 1 and there are nine trucks, how many cars are there?

(A) 3 (B) 9 (C) 12 (D) 27 [1]

8 If it costs $117 to clean the carpet in three rooms, what is the cost per room?

(A) $34 (B) $37 (C) $39 (D) $43 [1]

9 How many litres will flow into a drum in half an hour at the rate of 5 litres per minute?

(A) 25 (B) 75 (C) 150 (D) 250 [1]

10 If there are 80 boys and 60 girls, what is the ratio of girls to boys?

(A) 2 : 3 (B) 3 : 2 (C) 4 : 3 (D) 3 : 4 [1]

11 If you travel 240 km in 4 hours what is the average speed?

(A) 60 km/h (B) 70 km/h (C) 80 km/h (D) 90 km/h [1]

12 What is the ratio 45 minutes to an hour in simplest form?

(A) 45 : 1 (B) 4 : 5 (C) 2 : 3 (D) 3 : 4 [1]

13 Three-quarters of the counters in a box are black and the rest are white. What is the ratio of black to white counters, in simplest form?

(A) 3 : 4 (B) 3 : 1 (C) 4 : 1 (D) 1 : 4 [1]

14 If the fuel usage of a car is 8 L per 100 km, how far can it travel on 56 L?

(A) 700 km (B) 560 km (C) 800 km (D) 448 km [1]

15 How long will it take to travel 245 km at an average speed of 70 km/h?

(A) 3 h (B) $3\frac{1}{2}$h (C) $3\frac{3}{4}$h (D) 4 h [1]

Total marks /15

Ratio and Rates Non-calculator

LEVEL 2 TEST PART A

Time allowed: 40 minutes

	Marks

1 What is the ratio 4 : 6 : 2 in simplest form?
Ⓐ 3 : 5 : 1 Ⓑ 2 : 3 : 1 Ⓒ 1 : 3 : 5 Ⓓ 1 : 2 : 3 — 1

2 What is the ratio of $5 to 60 cents in simplest form?
Ⓐ 25 : 3 Ⓑ 5 : 6 Ⓒ 1 : 12 Ⓓ none of these — 1

3 The scale on a map is 1 : 100 000. This means that 1 cm represents:
Ⓐ 1 m Ⓑ 100 m Ⓒ 1 km Ⓓ 100 km — 1

4 What is the average speed of a train that travels 70 km in half an hour?
Ⓐ 35 km/h Ⓑ 105 km/h Ⓒ 120 km/h Ⓓ 140 km/h — 1

5 The ratio of adults to children at a show is 5 : 3. If there are 120 children, how many adults are there?
Ⓐ 72 Ⓑ 75 Ⓒ 175 Ⓓ 200 — 1

6 A fertiliser is applied at the rate of 0.1 kg/m^2. How many kilograms will be needed to cover an area of 25 m^2?
Ⓐ 0.5 Ⓑ 2.5 Ⓒ 6.25 Ⓓ 62.5 — 1

7 A boat is travelling at 55 km/h. How long will it take to cover a distance of 11 km?
Ⓐ 5 min Ⓑ 10 min Ⓒ 12 min Ⓓ 15 min — 1

8 A man walks 250 m in 3 minutes. What is his average speed?
Ⓐ 4 km/h Ⓑ 5 km/h Ⓒ $7\frac{1}{2}$ km/h Ⓓ $8\frac{1}{3}$ km/h — 1

9 Some pegs are in a basket. One-quarter are red, one-third are blue and the rest are white. What is the ratio of red to blue to white pegs?
Ⓐ 3 : 4 : 5 Ⓑ 3 : 4 : 1 Ⓒ 4 : 3 : 1 Ⓓ 4 : 3 : 2 — 1

10 A car uses petrol at the rate of 7.2L/100 km. How much petrol will it use to travel 500 km?
Ⓐ 36 L Ⓑ 14.4 L Ⓒ 144 L Ⓓ 78.8 L — 1

11 Water is flowing at 15 L per minute. How long will it take to fill a 3000 L tank?
Ⓐ 45 min Ⓑ 2 h Ⓒ 3 h 20 min Ⓓ $7\frac{1}{2}$h — 1

12 The ratio of men to women to children on a bus is 3 : 2 : 5. If there are 30 children, how many women are on the bus?
Ⓐ 27 Ⓑ 18 Ⓒ 12 Ⓓ 8 — 1

13 If it takes 20 minutes to travel 24 km this is a rate of:
Ⓐ 64 km/h Ⓑ 72 km/h Ⓒ 75 km/h Ⓓ 80 km/h — 1

14 The cost of a carpet is $60/m^2. What is the cost of 20 square metres of the carpet?
Ⓐ $1200 Ⓑ $7200 Ⓒ $24 000 Ⓓ $72 000 — 1

15 The ratio $\frac{1}{2} : \frac{1}{3} : \frac{1}{4}$ in simplest form is:
Ⓐ 2 : 3 : 4 Ⓑ 4 : 3 : 2 Ⓒ 6 : 3 : 4 Ⓓ 6 : 4 : 3 — 1

Ratio and Rates Non-calculator **(Level 2 Test)**

PART B Write answers only in the space provided.

16 If \$120 is divided in the ratio 5 : 7, what is the smaller amount? ______________ [1]

17 How far will you travel in $1\frac{1}{2}$ hours at 68 km/h? ______________ [1]

18 If a car uses fuel at the rate of 8 L/100 km, how far will it travel on 100 L? ______________ [1]

19 What is the ratio 8 : 12 : 4 in simplest form? ______________ [1]

20 The ratio of male to female students at a college is 5 : 6.
If there are 330 students at the college, how many are female? ______________ [1]

21 How long will it take, in hours and minutes, to travel 75 km at 60 km/h? ______________ [1]

22 Three-fifths of the cars in a car park are white. What is the ratio of white cars to all other cars? ________ [1]

23 The scale on a map is 1 : 500. What actual distance, in metres, will a length of 2.5 cm on the map represent? ______________ [1]

24 What is the average speed, in km/h, if a person runs 3 km in 15 minutes? ______________ [1]

25 The ratio of trees to shrubs in a park is 2 : 7. If there are 14 trees, how many shrubs are there? ________ [1]

PART C Show all working.

26 The ratio of boys to girls in a team is 3 : 2.

a If there are 15 people in the team, how many are boys? [1 mark]

____

b If 4 boys leave the team and are replaced by 4 girls, what is the new ratio of boys to girls? [2 marks]

____

____

27 A car left Birdwood at 8:35 am and travelled to Rayville, a distance of 150 km, arriving at 11:05 am.

a What was the average speed of the trip? [2 marks]

____

____

b If the car used petrol at the rate of 8 L/100 km, and petrol cost \$1.25 per litre, what was the cost of the petrol used on the trip? [2 marks]

____

____

c If the car averaged 75 km/h on the return journey, how long would it take? [1 mark]

____

28 A recipe for toffee uses 1 cup sugar and $\frac{1}{4}$ cup of water.

a What is the ratio of sugar to water, in simplest form? [1 mark]

____

b Rose uses a slightly different recipe. She uses water and vinegar in the ratio 3 : 1 instead of plain water. What is the ratio of sugar to water to vinegar in Rose's recipe? [2 marks]

____

____

29 Sadie's Solutions is paid \$180 per hour for contract work.

a How much would Sadie's Solutions be paid for 6 hours work? [1 mark]

____

b Sadie, Sam and Sarah share the money that Sadie's Solutions receives in the ratio 4 : 3 : 2. How much of each hour's pay does Sadie receive? [2 marks]

____

____

c How many hours would need to be worked so that Sadie received \$4000? [1 mark]

____

Total marks /40

LEVELS 1 AND 2 TEST PART A

Time allowed: 40 minutes

Marks

1 What is the ratio 57 : 36 in simplest form?
(A) 21 : 1 (B) 19 : 12 (C) 12 : 7 (D) 14 : 9 [1]

2 Travelling 651 km in 7 hours 45 minutes is what average speed?
(A) 84 km/h (B) 81 km/h (C) 78 km/h (D) 87 km/h [1]

3 The scale on a plan is 1 : 250. A length of 5 cm on the plan would represent what real distance?
(A) 50 m (B) 12.5 m (C) 5 m (D) 1.25 m [1]

4 The ratio of old to new coins in a collection is 11 to 8. If there are 1672 old coins, what is the total number of coins in the collection?
(A) 1216 (B) 2888 (C) 2292 (D) 3971 [1]

5 In a cricket match the team batting second needed to score 275 runs in 50 overs. After 30 overs they had scored 141 runs. At what rate of runs per over do they need to score in the last 20 overs?
(A) 4.7 (B) 5.5 (C) 6.7 (D) 7.05 [1]

6 In simplest form the ratio 25 minutes to 2 hours is:
(A) 25 : 2 (B) 1 : 8 (C) 1 : 12 (D) 5 : 24 [1]

7 How far will you travel in an hour and a quarter at 60 km/h?
(A) 62.5 km (B) 75 km (C) 85 km (D) 90 km [1]

8 What is the ratio 72 : 48 : 60 in simplest form?
(A) 9 : 6 : 7 (B) 8 : 5 : 6 (C) 4 : 2 : 3 (D) 6 : 4 : 5 [1]

9 Alex and Max share \$26 000 in the ratio 3 : 2. What is Max's share?
(A) \$13 000 (B) \$12 200 (C) \$10 400 (D) \$9600 [1]

10 A mixture has 250 mL of bleach in 7.5 L of water. What is the ratio of bleach to water?
(A) 10 : 3 (B) 1 : 3 (C) 1 : 30 (D) 1 : 300 [1]

11 A plane flies 832 km at an average speed of 260 km/h. How long does the flight take?
(A) 3 h 12 min (B) 3 h 20 min (C) 4 h 30 min (D) 4 h 55 min [1]

12 What is the ratio $2\frac{3}{4} : 4\frac{1}{3}$ in simplest form?
(A) 11 : 13 (B) 39 : 44 (C) 33 : 52 (D) 7 : 10 [1]

13 A real distance of 125 m is represented on a map by a distance of 2.5 cm. What scale is used?
(A) 1 : 200 (B) 1 : 500 (C) 1 : 2000 (D) 1 : 5000 [1]

14 What is the ratio of 20 cents to 5 dollars in simplest form?
(A) 4 : 1 (B) 1 : 4 (C) 1 : 20 (D) 1 : 25 [1]

15 The ratio of men to women to children is 9 : 11 : 12. If there are 12 more women than men, how many children are there?
(A) 72 (B) 108 (C) 132 (D) not enough information [1]

PART B Write answers only in the space provided.

16 What is the ratio 42 : 30 : 48 in simplest form? ________________ [1]

17 585 km in 9 hours is what simple rate? ________________ [1]

18 If Abdul and Ali share $12 000 in the ratio 2 : 3, what is Ali's share? ________________ [1]

19 At 68 km/h, how long, in hours and minutes, will it take to travel 221 km? ________________ [1]

20 The ratio of pigs to sheep on a farm is 5 : 3. If there are 420 pigs, how many sheep are there? ________ [1]

21 What is the ratio of $2\frac{1}{2}$ hours to 20 minutes in simplest form? ________________ [1]

22 How much lime is needed to spread on 3.2 ha at the rate of 1.5 t/ha? ________________ [1]

23 If a heart beats 65 times an hour, how many times will it beat in a day? ________________ [1]

24 If $\frac{5}{8}$ of the people at a concert are female, what is the ratio of males to females at the concert? ________________ [1]

25 Travelling at 72 km/h, how far will you go in 3 h 40 min? ________________ [1]

PART C Show all working.

26 Briefly describe the difference between a ratio and a rate. [1 mark]

____

27 A scale drawing has been made of a block of land.

a The western boundary of the land is 24 m long and appears on the drawing as 6 cm long. What is the scale used, expressed as a ratio in simplest form? [1 mark]

____

b The northern boundary is 34 m long. How long is it on the diagram? [1 mark]

____

c The southern boundary is 13.5 cm long on the diagram. How long is the actual boundary? [1 mark]

____

28 When Jo left home, the odometer reading on her car was 87 649 km. When she reached her destination after $7\frac{1}{4}$ hours of actual driving, it read 88 171 km.

a How many kilometres did Jo drive? [1 mark]

____

b What was the average speed? [1 mark]

____

c Jo filled the car with petrol immediately before she left home and again when she arrived. If she paid $54.81 for the fuel at $1.40 per litre, how much petrol did Jo use? [1 mark]

____

d What was the rate of petrol consumption, in L/100 km? [1 mark]

____

29 a If 250 mL of water flows through a pipe in 4 seconds, what is the rate in litres per minute? [1 mark]

____

b How much water will flow through the pipe in an hour and a half? [1 mark]

____

c How long, in days, hours and minutes, would it take for 24 000 litres to flow through the pipe? [2 marks]

____

____

30 Grass seed is being planted on a paddock at the rate of 8 kg/ha. The seed is a combination of rye, chicory and clover in the ratio 10 : 3 : 2 by weight. What weight of each type of seed would be planted on a 4.5 ha paddock? [3 marks]

____

____

____

Total marks /40

Ratio and Rates Calculator Allowed

LEVELS 2 AND 3 TEST — PART A

Time allowed: 40 minutes

Marks

1 How long will it take to travel 256 km at 80 km/h?

(A) 3 h 2 min (B) 3 h 12 min (C) 3 h 16 min (D) 3 h 20 min — 1

2 What is the ratio 28.8 : 36 : 21.6 in simplest form?

(A) 4 : 5 : 3 (B) 5 : 6 : 4 (C) 7 : 9 : 6 (D) 8 : 10 : 7 — 1

3 A helicopter flies at 210 km/h. How far will it travel in 25 minutes?

(A) 52.5 km (B) 70 km (C) 87.5 km (D) 100 km — 1

4 Stacey, Nat and Chad share $6600 in the ratio 2 : 5 : 4. How much is Chad's share?

(A) $2200 (B) $2500 (C) $2400 (D) $3000 — 1

5 Which is equivalent to $2\frac{1}{4} : 3\frac{2}{3}$?

(A) 2 : 3 (B) 9 : 11 (C) 11 : 12 (D) 27 : 44 — 1

6 Spray is being applied to a lawn at the rate of 0.15 kg/m^2. How many kilograms of spray are needed to cover 24 square metres?

(A) 36 (B) 3.6 (C) 5.4 (D) 86.4 — 1

7 A bag holds red, blue, green and yellow marbles in the ratio 6 : 4 : 2 : 3. If there are 180 marbles altogether, how many more red than blue marbles are there?

(A) 24 (B) 12 (C) 36 (D) 15 — 1

8 Mel sets off at 8:45 am and drives for 65 km, arriving at 9:35 am. What is Mel's average speed?

(A) 60 km/h (B) 65 km/h (C) 72 km/h (D) 78 km/h — 1

9 A carwash product says to use half a cap to 5 L of water. If a cap holds 120 mL of the product, how much is needed for each litre of water?

(A) 6 mL (B) 12 mL (C) 30 mL (D) 60 mL — 1

10 What is the ratio of 120 g to 3 kg in simplest form?

(A) 40 : 1 (B) 1 : 40 (C) 1 : 25 (D) 25 : 1 — 1

11 If the ratio of dogs to cats is 7 : 5 and the ratio of dogs to guinea pigs is 9 : 2, what is the ratio of cats to guinea pigs?

(A) 5 : 2 (B) 7 : 2 (C) 35 : 18 (D) 45 : 14 — 1

12 Some land on a property is cleared and the rest is forested. The ratio of cleared land to forested land is 5 : 2. If there are 120 ha of forested land, how many hectares are there altogether?

(A) 420 (B) 300 (C) 288 (D) 168 — 1

13 One-third of the pegs in a box are pink, two-fifths are yellow and the rest are blue. What is the ratio of pink to yellow to blue pegs?

(A) 5 : 4 : 6 (B) 4 : 5 : 6 (C) 5 : 6 : 4 (D) 6 : 5 : 4 — 1

14 There were 18 boys and 15 girls on an oval. How many girls left, if the ratio of boys to girls is now 3 : 2 and the number of boys hasn't changed?

(A) 2 (B) 3 (C) 4 (D) 5 — 1

15 Which is closest to the amount of time needed to fill a swimming pool that holds 45 000 L, if water flows into the pool from a hose at 5 L per minute?

(A) 15 h (B) 3 h (C) 3 days (D) 6 days — 1

Ratio and Rates Calculator Allowed (Levels 2 and 3 Test)

PART B Write answers only in the space provided.

16 What is the ratio of students to teachers if there are 33 teachers and 594 students? ______ [1]

17 How long, in hours and minutes, will it take to travel 216 km at 64 km/h? ______ [1]

18 What is the ratio of $2\frac{1}{2}$ hours to 15 minutes in simplest form? ______ [1]

19 If the ratio of boys to girls is 3 : 8 and there are 528 girls, how many boys? ______ [1]

20 What is the ratio of 25 cm to 1.2 m in simplest form? ______ [1]

21 What is the average weight of an apple in grams, if 12 apples were bought for $6.72 and the apples were priced at $3.50/kg? ______ [1]

22 If $684 is divided in the ratio 3 : 5 : 1, what is the largest share? ______ [1]

23 What is the new price of a fridge if it was priced at $888, but the price has increased in the ratio 5 : 4? ______ [1]

24 The ratio of red to blue to green tickets was 7 : 5 : 3. If there were 81 green tickets, how many more red than blue tickets were there? ______ [1]

25 If two-thirds of the days of a holiday were fine, what was the ratio of wet to fine days? ______ [1]

PART C Show all working.

26 A map has a scale of 1 : 100 000.

a The distance from Ashton to Bradden on the map is 3.5 cm. What would the actual distance between those towns be? [1 mark]

______

b The actual straight-line distance from Pindale to Queenville is 17.5 km. How long is the distance from Pindale to Queenville on the map? [1 mark]

______

c What is the ratio of the distance between Pindale and Queenville to the distance from Ashton to Bradden in simplest form? [1 mark]

______

d If a helicopter flew from Pindale to Queenville in 5 minutes, what is its speed in kilometres per hour? [1 mark]

______

27 Instructions for a lawn additive say to use 7.5 g per square metre.

a How many square metres will 45 g cover? [1 mark]

______

b The additive needs to be added to water. The amount of water required is 120 mL/g. How many square metres would be covered with 9 L of water? [2 marks]

______

______

28 Robert ran a marathon, a distance of 42.195 km, in 2 hours 25 minutes.

a What was Robert's average speed? [1 mark]

______

b Steve ran the same marathon. His average speed was 14.55 km/h. How long, in hours and minutes, did Steve take to complete the marathon? [1 mark]

______

c About how many more kilometres did Steve still have to run at the time Robert finished? [2 marks]

______

29 To make concrete, Lisa needs cement, sand and aggregate (metal) in the ratio 1 : 2 : 3 by volume.

a Sixteen 20 kg bags of cement are needed to make one cubic metre of concrete. Lisa needs 36 bags. How many cubic metres of concrete does she intend to make? [1 mark]

______

b How much sand and aggregate does Lisa need? [2 marks]

______

______

c How many cubic metres of concrete does each kilogram of cement make? [1 mark]

______

Total marks ___ / 40

Mixed Number Non-calculator

LEVEL 2 TEST — PART A

Time allowed: 40 minutes

Marks

1 $-2 \times -3 \times -4 =$

(A) -24 (B) 24 (C) -9 (D) 9 — 1

2 $7^0 =$

(A) 0 (B) 1 (C) 7 (D) 8 — 1

3 $0.2\dot{1}\dot{5}$

(A) 0.215 555 555…
(B) 0.215 151 515 15…
(C) 0.215 215 215…
(D) 0.215 152 1515… — 1

4 \$500 increased by 20% is:

(A) \$520 (B) \$550 (C) \$600 (D) \$700 — 1

5 The ratio of dogs to cats seen by a vet is 5 : 3. If there are 120 cats, how many dogs are there?

(A) 200 (B) 75 (C) 320 (D) 172 — 1

6 How long does it take a car to travel 140 km at an average speed of 60 km/h?

(A) 2 h 10 min (B) $2\frac{1}{4}$ h (C) 2 h 20 min (D) $2\frac{1}{2}$ h — 1

7 Which of these is an irrational number?

(A) $\frac{5}{3}$ (B) $\sqrt{6}$ (C) $0.\dot{1}4285\dot{7}$ (D) 350% — 1

8 $5 - 6 \times 3 =$

(A) -3 (B) 3 (C) -13 (D) 13 — 1

9 $2^4 \times 2^3 =$

(A) 2^7 (B) 2^{12} (C) 4^7 (D) 4^{12} — 1

10 What is the profit as a percentage of the cost price if something bought for \$25 is sold for \$30?

(A) 5% (B) 10% (C) 20% (D) 25% — 1

11 If \$600 is divided in the ratio 3 : 5 : 2, what is the smallest share?

(A) \$20 (B) \$60 (C) \$100 (D) \$120 — 1

12 $-12 \div 2 \times -3 =$

(A) -2 (B) 2 (C) -18 (D) 18 — 1

13 In index form, $(3^2)^5 =$

(A) 3^7 (B) 3^{10} (C) 9^7 (D) 9^{10} — 1

14 $-\frac{2}{3} \times -\frac{3}{5} =$

(A) $\frac{2}{5}$ (B) $-\frac{2}{5}$ (C) $-\frac{1}{3}$ (D) $\frac{1}{3}$ — 1

15 $5^9 \div 5^3 \div 5 =$

(A) 1 (B) 5^2 (C) 5^3 (D) 5^5 — 1

PART B Write answers only in the space provided.

16 What is $0.\dot{6}$ correct to three decimal places? _______________ 1

17 What is the result if \$400 is decreased by 10%? _______________ 1

18 $-2 \times -5 - 3 \times -4 =$ _______________ 1

19 $0.8 \div -0.2 =$ _______________ 1

20 Evaluate $3^0 + 3^3$. _______________ 1

21 Simplify the ratio \$3 to \$1.50 to 25 cents. _______________ 1

22 $\frac{-12+16}{8--4} =$ _______________ 1

23 How far will a vehicle travel in 40 minutes at an average speed of 90 km/h? _______________ 1

24 An increase from \$10 to \$25 is what percentage increase? _______________ 1

25 What is $(2^6 \times 2^2)^3$ in simplest index form? _______________ 1

PART C Show all working.

26 Given that $67 \times 32 = 2144$ and $2144 \times 19 = 40\,736$, find the value of:

a $-67 \times -32 \times -19$ (1 mark)

____

b $-40\,736 \div -19$ (1 mark)

____

c $2144 \div -67$ (1 mark)

____

27 Express $\frac{7}{12}$ as a recurring decimal. (2 marks)

____

____

28 a How much is a discount of 20% on clothes priced at \$160? (1 mark)

____

b A suit is discounted from \$400 to \$340. What percentage discount is that? (1 mark)

____

c The shop makes a profit of 200% on scarves that sell for \$6 each. What does the shop pay for the scarves? (1 mark)

____

29 The ratio of dolls to teddy bears in a collection is 4 : 3.

a If there are 60 teddy bears, how many dolls are there? (1 mark)

____

b Sixteen of the dolls and 12 of the teddy bears are sold. Does the ratio of dolls to teddy bears change? Explain why or why not. (2 marks)

____

____

30 A car travels 420 km in 6 hours.

a What is the average speed in kilometres per hour? (1 mark)

____

b If the car uses petrol at the rate of 8 L/100 km, how much petrol will it have used? (1 mark)

____

31 Shirley gave the answer to $7^3 \times 7^4$ as 49^7. Was Shirley correct? Explain why or why not. (2 marks)

____

____

____

Total marks /40

Mixed Number Calculator Allowed

LEVEL 2 TEST PART A

Time allowed: 40 minutes

1 $\frac{5}{22}$ as a decimal is:

(A) $0.22\dot{7}$ (B) $0.2\dot{2}\dot{7}$ (C) $0.\dot{2}2\dot{7}$ (D) $0.\dot{2}\dot{7}$ | Marks: 1

2 \$912 increased by 35% is:

(A) \$1231.20 (B) \$1319.20 (C) \$947 (D) \$2143.20 | 1

3 What is the ratio, in simplest form, of 5 minutes to $2\frac{1}{4}$ hours?

(A) 9 : 20 (B) 1 : 43 (C) 1 : 27 (D) 1 : 29 | 1

4 Which of these is rational?

(A) $\sqrt{189}$ (B) $\sqrt{289}$ (C) $\sqrt{389}$ (D) $\sqrt{489}$ | 1

5 Travelling 98 km in 1 hour 45 minutes gives what average speed?

(A) 56 km/h (B) 68 km/h (C) 99 km/h (D) 172 km/h | 1

6 $-16 - 5 \times -9 =$

(A) 99 (B) 189 (C) –61 (D) 29 | 1

7 What was the buying price if a car was sold for \$12 000, which represented a loss of 25%?

(A) \$21 000 (B) \$15 000 (C) \$18 000 (D) \$16 000 | 1

8 $(5^3)^9 =$

(A) 5^{12} (B) 5^{27} (C) 25^{27} (D) 125^{12} | 1

9 A box holds apples, oranges and lemons in the ratio 9 : 5 : 2. There are 10 lemons, so how many more apples than oranges are there?

(A) 4 (B) 15 (C) 20 (D) 25 | 1

10 $(-2)^0 =$

(A) 0 (B) 1 (C) –1 (D) –2 | 1

11 What is $\frac{8}{37}$ correct to five decimal places?

(A) 0.216 22 (B) 0.216 16 (C) 0.2162 (D) 0.216 17 | 1

12 $\frac{-3-9}{-3+6} =$

(A) 6 (B) 2 (C) $1\frac{1}{3}$ (D) –4 | 1

13 A book was bought for \$33. The price included GST of 10%. How much was the GST?

(A) \$30 (B) \$2.97 (C) \$3.00 (D) \$3.30 | 1

14 Which statement is correct?
I All recurring decimals are non-terminating.
II All non-terminating decimals are recurring.

(A) I but not II (B) II but not I (C) both I and II (D) neither I nor II | 1

15 Which numbers are in ascending order?

(A) 2.3, $\frac{7}{3}$, 203% (B) $\frac{7}{3}$, 2.3, 203% (C) 203%, 2.3, $\frac{7}{3}$ (D) 203%, $\frac{7}{3}$, 2.3 | 1

PART B Write answers only in the space provided.

16 What is the value of π to four decimal places? ________________ 1

17 Write $0.\dot{7}\dot{5}$ as a fraction. ________________ 1

18 An increase from \$400 to \$1000 is an increase of what per cent? ________________ 1

19 What is the ratio $\frac{1}{3} : \frac{1}{4} : \frac{1}{5}$ in simplest form? ________________ 1

20 $-18 - -72 \div -6 \times -4 =$ ________________ 1

21 What is the percentage profit if a ring bought for \$2500 is sold for \$3250? ________________ 1

22 How long, in hours and minutes, does it take to travel 255 km at 75 km/h? ________________ 1

23 What is $7^{12} \div 7^4 \times 7^3$ in simplest index form? ________________ 1

24 $-0.2 \times 0.7 - 0.4 \times -0.5 =$ ________________ 1

25 What is the result if 600 g is increased by 120%? ________________ 1

PART C Show all working.

26 Water flows from a tank of capacity 1940 L, at the rate of 1 L every 3 seconds.

a What is the flow rate in litres per minute? 1 mark

____

b How long, in hours and minutes, will it take the tank to empty if it was full to begin with? 1 mark

____

c What is the answer to part **b** in hours? Give the answer as a decimal. 1 mark

____

27 a What is the percentage increase if \$450 is increased by \$54? 1 mark

____

b If \$3450 is increased by the same percentage, what is the new amount? 1 mark

____

c What amount would increase to \$1792 at the same percentage? 1 mark

____

28 What is the ratio $0.\dot{6} : 0.6\dot{3}$ in simplest form? 2 marks

____

____

29 The minimum temperatures, in °C, at a camp last week are shown in the table.

Day	Mon.	Tue.	Wed.	Thu.	Fri.
Temperature	–2	3	–5	–7	–4

a The difference between the maximum and minimum temperatures on Friday was 26°. What was the maximum temperature? 1 mark

____

b What was the average minimum for the five days? 2 marks

____

____

30 A coat was bought for \$350 and sold at a profit of 15%.

a What was the selling price? 1 mark

____

b What was the ratio of cost price to selling price in simplest form? 1 mark

____

c A tie was bought for \$25 and sold at a profit of 115%. What was the selling price of the tie? 1 mark

____

d What is the ratio of the cost price to selling price of the tie, in simplest form? 1 mark

____

Total marks /40

PRETEST

Time allowed: 15 minutes **Total marks: 15**

					Marks
1	$a + a + a + a + a =$				
	(A) $a + 5$	(B) $5a$	(C) a^5	(D) none of these	1
2	$3a + 4a =$				
	(A) $7a$	(B) $7a^2$	(C) $12a$	(D) $12a^2$	1
3	If $x = 3$, then $x + 6 =$				
	(A) 3	(B) 9	(C) 18	(D) 36	1
4	$3a + 5b =$				
	(A) $8ab$	(B) $15ab$	(C) $35ab$	(D) cannot be simplified	1
5	$3a \times 5b =$				
	(A) $8ab$	(B) $15ab$	(C) $35ab$	(D) cannot be simplified	1
6	$2x \times 4x =$				
	(A) $6x$	(B) $8x$	(C) $6x^2$	(D) $8x^2$	1
7	$p + 4 =$				
	(A) t	(B) $4p$	(C) $4 + p$	(D) p^4	1
8	$9k - 3k =$				
	(A) 3	(B) $3k$	(C) 6	(D) $6k$	1
9	$7n \div 7 =$				
	(A) 0	(B) 1	(C) 7	(D) n	1
10	If $q = 5$, what is $3q$?				
	(A) 8	(B) 15	(C) 35	(D) none of these	1
11	$6a \div 3a =$				
	(A) 2	(B) 3	(C) $2a$	(D) $3a$	1
12	If $x = 8$ and $y = 4$ then $xy =$				
	(A) 2	(B) 12	(C) 32	(D) 84	1
13	$6m \div m =$				
	(A) 0	(B) 1	(C) 6	(D) m	1
14	The number 5 less than x is:				
	(A) $x - 5$	(B) $5 - x$	(C) $5x$	(D) s	1
15	$\frac{8a}{4b} =$				
	(A) $\frac{2a}{b}$	(B) $\frac{4a}{b}$	(C) $\frac{a}{2b}$	(D) $\frac{a}{4b}$	1

Total marks /15

LEVEL 1 TEST PART A

Time allowed: 40 minutes

Marks

1 $5p + 3q + 2p =$

(A) $10pq$ (B) $7p + 3q$ (C) $10(p + q)$ (D) cannot be simplified — 1

2 $3x \times 2y \times 4z =$

(A) $9xyz$ (B) $24xyz$ (C) $24(x + y + z)$ (D) cannot be simplified — 1

3 $3x + 2y + 4z =$

(A) $9xyz$ (B) $24xyz$ (C) $24(x + y + z)$ (D) cannot be simplified — 1

4 If $m = 3$ and $n = 5$ then $4mn =$

(A) 435 (B) 240 (C) 60 (D) 12 — 1

5 $9abc \div 3b =$

(A) $3ac$ (B) $3ab^2c$ (C) $6ac$ (D) $6ab^2c$ — 1

6 $6x^2 + 2x^2 =$

(A) $40x^4$ (B) $40x^2$ (C) $8x^4$ (D) $8x^2$ — 1

7 $7y + 5y - 9y =$

(A) 3 (B) $3y$ (C) $-3y$ (D) cannot be simplified — 1

8 $3x \times 2x + 4x =$

(A) $10x$ (B) $10x^3$ (C) $18x^2$ (D) $6x^2 + 4x$ — 1

9 If $z = 3$, what is $2z^2$?

(A) 18 (B) 29 (C) 36 (D) 529 — 1

10 $12a - 5a + 6a =$

(A) 1 (B) a (C) 13 (D) $13a$ — 1

11 $12abc \div 6ac =$

(A) $2b$ (B) $6b$ (C) $2ac$ (D) $6ac$ — 1

12 $4t + 3u + 6t + u =$

(A) $10t + 3u$ (B) $10t + 3u^2$ (C) $10t + 4u$ (D) $13t^2u^2$ — 1

13 $5 \times 3a + 6 \times 2a =$

(A) $27a^2$ (B) $115a^2$ (C) $27a$ (D) $42a$ — 1

14 $5a + 2b - 3a + b =$

(A) $2a + 3b$ (B) $2a + b$ (C) $8a + b$ (D) $8a + 3b$ — 1

15 $18x \div 3x \times 2x =$

(A) $12x^2$ (B) $12x$ (C) $3x$ (D) $\frac{3}{x}$ — 1

PART B Write answers only in the space provided.

16 $7x + 3x - 4x =$ ______________ [1]

17 $-5b + 6b =$ ______________ [1]

18 $5x \times 4x =$ ______________ [1]

19 $6 - 2m + 3 - 4m =$ ______________ [1]

20 $k \times 11k =$ ______________ [1]

21 $3p + 4 \times 2p =$ ______________ [1]

22 $35ab \div 7b =$ ______________ [1]

23 $6x + 5y - 4x + 3y =$ ______________ [1]

24 $8a \times -10b =$ ______________ [1]

25 $9x \times 2 \times 5x =$ ______________ [1]

PART C Show all working.

26 If $a = 5$, $b = 8$ and $c = 4$, find the value of:

a $4a + 2b + 3c$ [1 mark]

____

b $b^2 - ac$ [1 mark]

____

c $\frac{ab}{c}$ [1 mark]

____

27 Simplify:

a $3x \times (2x + 5x) - 4x \times 5x$ [2 marks]

____

____

b $6a + 3b + 4 - 5a + 7b - 3 + b$ [2 marks]

____

____

c $\frac{8m \times 5m}{4m^2 \div 2m}$ [2 marks]

____

____

d $9n - n + 7n^2 - n^2 + 6n - 3n^2$ [2 marks]

____

____

e $7 \times \frac{4a \times 3b}{6ab}$ [2 marks]

____

____

f $9p^2 + 2p \times 7p - 16p^2 \div 4p$ [2 marks]

____

____

Total marks /40

LEVEL 2 TEST — PART A

Time allowed: 40 minutes

Marks

1 $m^2 + m^2 =$

(A) m^4 (B) $2m^2$ (C) $2m^4$ (D) none of these — 1

2 $5a - 3b + 4a - 6b =$

(A) $a - 9b$ (B) $a + 9b$ (C) $9a - 9b$ (D) $9a + 9b$ — 1

3 $7a + 5 \times 3a =$

(A) $22a$ (B) $36a$ (C) $36a^2$ (D) $21a^2 + 15a$ — 1

4 If $u = -3$, then $2u^2 =$

(A) -18 (B) -36 (C) 18 (D) 36 — 1

5 $6p + 3q - 5p + 4q =$

(A) $p + 7q$ (B) $p - q$ (C) $11p - q$ (D) $7q$ — 1

6 $9x^2 \div x^2 \times 9 =$

(A) 1 (B) 81 (C) x^2 (D) $81x^2$ — 1

7 $\dfrac{12a^2}{24a} =$

(A) $\dfrac{a}{2}$ (B) $\dfrac{1}{2a}$ (C) $\dfrac{2}{a}$ (D) $2a$ — 1

8 $7 - 6t - 11 + t =$

(A) $18 - 5t$ (B) $-5t - 4$ (C) $-7t - 4$ (D) $4 - 7t$ — 1

9 If $a = 5$ and $b = -3$, then $4ab - 3b =$

(A) -69 (B) -66 (C) -54 (D) -51 — 1

10 $8x - 2x \times 5x =$

(A) $30x$ (B) $30x^2$ (C) $-2x^2$ (D) $8x - 10x^2$ — 1

11 Which is the same as $\dfrac{p}{qr}$?

(A) $p \div (q \times r)$ (B) $p \div q \times r$ (C) $\dfrac{p}{q} \times r$ (D) all of these — 1

12 $3a \times 9b - 7a \times 4b =$

(A) 0 (B) $-ab$ (C) ab (D) $24ab$ — 1

13 $-4n^2 + 7n - n^2 - 2n =$

(A) $-3n^2 + 5n$ (B) $-3n^2 - 5n$ (C) $5n - 5n^2$ (D) $-5n^2 - 5n$ — 1

14 What number is 6 less than the square of x?

(A) $6 - x^2$ (B) $6 - \sqrt{x}$ (C) $x^2 - 6$ (D) $\sqrt{x} - 6$ — 1

15 $36x \div 9x \times 4x =$

(A) 1 (B) x (C) $\dfrac{1}{x}$ (D) $16x$ — 1

PART B Write answers only in the space provided.

16 $-3x^2 + 7x^2 =$ ______ [1]

17 $5m^2 + 3m \times 2m =$ ______ [1]

18 $18x^2yz^2 \div 6xy =$ ______ [1]

19 $5m - 4n + 7m + n =$ ______ [1]

20 $13a^2 - 7a - 9a^2 + 11a =$ ______ [1]

21 $-4ac \times 3ab =$ ______ [1]

22 $48d^2 \div 8d \times 6d =$ ______ [1]

23 If $h = -2$, what is the value of $5 - 3h$? ______ [1]

24 Simplify $\frac{4ab}{12a^2}$. ______ [1]

25 $-5ab \times -2ac =$ ______ [1]

PART C Show all working.

26 Find the value of $-3x^2 + 5x - 4x^2 + 8x + 7x^2 - 9x$ when $x = 3$. [2 marks]

27 Simplify:

a $(4x + 3x) \times (12x - 5x) - x^2$ [2 marks]

b $36abc \div 9ac \times 2bc + 10ab$ [2 marks]

c $6m - 8n - 4m + 5n - 7 + 2m - 9 + m$ [2 marks]

d $\frac{13k^2 + 5k \times 7k}{6k \times 4k}$ [2 marks]

e $\frac{-v \times 6 + 3v \times -4}{9v^2 \div v}$ [2 marks]

28 If $p = -2$, $q = 5$ and $r = -1$, find the value of:

a $pq - r$ [1 mark]

b $4p^2$ [1 mark]

c $\frac{6qr}{15p}$ [1 mark]

Total marks /40

LEVEL 3 TEST PART A

Time allowed: 40 minutes

Marks

1 If $a = -2$ and $b = -3$, then $ab^2 =$

(A) 18 (B) 36 (C) –18 (D) –36 [1]

2 $2x^2 + 5x - 7x^2 + 3x + 4x^2 - 6x =$

(A) $2x - x^2$ (B) $13x^2 + 8x$ (C) $-x^2 + 8x$ (D) $13x^2 + 2x$ [1]

3 $-4a \times 2a - 3a \times -5a =$

(A) $-23a^2$ (B) $7a^2$ (C) $25a^2$ (D) $-20a^2$ [1]

4 $5p + 2 \times 3p - p \times 4 + 6p \times -2$

(A) $-46p$ (B) $19p$ (C) $-5p$ (D) $9p$ [1]

5 $\dfrac{9x^2 \div 9}{9x^2 \div x} =$

(A) $\dfrac{9}{x}$ (B) $\dfrac{x}{9}$ (C) $9x$ (D) $\dfrac{1}{9x}$ [1]

6 $9p - p - 4q + 3q =$

(A) $9 - 7q$ (B) $8p - 7q$ (C) $8p - q$ (D) $9 - q$ [1]

7 The number 7 less than $2x - 5$ is:

(A) $2x - 12$ (B) $12 - 2x$ (C) $2x - 2$ (D) $2 - 2x$ [1]

8 $\dfrac{12abc^2}{3ab^2c} =$

(A) $4abc$ (B) $4bc$ (C) $\dfrac{4b}{c}$ (D) $\dfrac{4c}{b}$ [1]

9 $28x^2 \div 7x \times 5x =$

(A) $\dfrac{4x^2}{5}$ (B) $\dfrac{4}{5}$ (C) $1\frac{1}{4}$ (D) $20x^2$ [1]

10 $-4a - 2b + 5a - b =$

(A) $-9a - b$ (B) $-9a - 3b$ (C) $a + 3b$ (D) $a - 3b$ [1]

11 If $p = -2$, $q = 7$ and $r = -5$, then $r - pq =$

(A) 9 (B) –19 (C) –21 (D) –49 [1]

12 $5a^2 - 3a + 8a + 4a^2 =$

(A) $9a^2 - 11a$ (B) $9a^2 + 5a$ (C) $9a^4 - 11a$ (D) $9a^4 + 5a$ [1]

13 $(-5m + 2m) \times (-3m - 4m) =$

(A) $-49m^2$ (B) $-21m^2$ (C) $49m^2$ (D) $21m^2$ [1]

14 $\dfrac{6x^2 - 8x^2}{4x \times 2x} =$

(A) $-\dfrac{1}{4}$ (B) x^2 (C) $2x^2$ (D) –6 [1]

15 $\dfrac{3x^2}{x} - x =$

(A) $3x$ (B) $2x$ (C) $3x - 1$ (D) $2x - 1$ [1]

PART B Write answers only in the space provided.

16 $\frac{1}{2}$ of $4p^2$ = ______________ [1]

17 $-2x^2 + 5x - 3x^2 + 8x + x^2$ = ______________ [1]

18 $15a \times -3b \div 5b$ = ______________ [1]

19 $4xy + 8x \times 3y - 6x + y$ = ______________ [1]

20 If $m = -4$ and $n = -2$, then $m^2 - n^2$ = ______________ [1]

21 $(7a + 4a) \times (6a - a)$ = ______________ [1]

22 $9y \div 3y + 8z \div 4z$ = ______________ [1]

23 $-5x^2 + 2x \times 6x - x^2$ = ______________ [1]

24 $(-3t + 5t) \times 7 - t + t^2$ = ______________ [1]

25 $-8 - 7d + 3d - -2d + (-6d) + 5$ = ______________ [1]

PART C Show all working.

26 Write expressions for each of these, simplifying where possible:

a the sum of $2x$ and $3y$ [1 mark]

____

b the product of $4a$ and $5b$ [1 mark]

____

c the number of cents in x dollars [1 mark]

____

27 Find the value of $5q^2 - 3q - 7q^2 + 6q + 2q^2 - 3$ when $q = -1$. [2 marks]

____

____

28 Simplify:

a $\dfrac{5x + 3x \times -7}{72x^2 \div 9x}$ [2 marks]

____

____

b $[(9a + 5a) \div 2a \times a] \times -2ab$ [2 marks]

____

____

c $\dfrac{24x^2 - 18x^2}{24x^2 \div 8x}$ [2 marks]

____

____

d $24pqr \div (-q + 4q) \times (-2pr + 3p \times -3r)$ [2 marks]

____

____

29 When asked to find a^2 when $a = -15$, Nell put -15^2 on her calculator and got -225. What should the answer have been? Briefly explain why Nell's answer was wrong. [2 marks]

____

____

Total marks /40

PRETEST

Time allowed: 15 minutes **Total marks: 16**

					Marks
1	Which is **not** a factor of 12?				
	(A) 4	(B) 6	(C) 8	(D) 12	1
2	$3 \times x \times 2 \times y =$				
	(A) $6xy$	(B) $6x + 6y$	(C) $5xy$	(D) $5 + x + y$	1
3	$4x$ is the same as:				
	(A) $4 + x$	(B) $x \times 4$	(C) $4 \div x$	(D) $40 + x$	1
4	Which of these is a factor of 36?				
	(A) 14	(B) 16	(C) 18	(D) 20	1
5	What is the product of 3 and $2x$?				
	(A) $5x$	(B) $6x$	(C) $5x^2$	(D) $6x^2$	1
6	$5a \times 3a =$				
	(A) $8a$	(B) $15a$	(C) $8a^2$	(D) $15a^2$	1
7	Which is **not** equal to $6ab$?				
	(A) $6a \times b$	(B) $3a + 3b$	(C) $3a \times 2b$	(D) $a \times 6b$	1
8	Which is a common factor of 20 and 35?				
	(A) 5	(B) 7	(C) 70	(D) 140	1
9	$8 \times (4 + 5) =$				
	(A) 37	(B) 44	(C) 72	(D) 160	1
10	$7x + 5 =$				
	(A) $12x$	(B) $35x$	(C) $75x$	(D) cannot be simplified	1
11	Which is **not** a factor of $8x^2$?				
	(A) $4x$	(B) 2	(C) $8x^2$	(D) $16x$	1
12	$9 \times 4 + 9 \times 7 =$				
	(A) 9×11	(B) 9×28	(C) 18×11	(D) 18×28	1
13	What is the highest common factor of 18 and 24?				
	(A) 2	(B) 6	(C) 9	(D) 12	1
14	What is the lowest common multiple of 18 and 24?				
	(A) 12	(B) 36	(C) 72	(D) 144	1
15	$9x \times 2x + 9x \times 5y =$				
	(A) $90x^2y$	(B) $180x^3y$	(C) $18x + 45xy$	(D) $18x^2 + 45xy$	1
16	$7 \times 2x + 7 \times 5 + 6 \times 2x + 6 \times 5 =$				
	(A) $26x + 65$	(B) $26x + 325$	(C) $52x + 65$	(D) $52x + 325$	1

Total marks /16

Expanding and Factorising Calculator Allowed

LEVEL 1 TEST PART A

Time allowed: 40 minutes

					Marks
1	Which is a factor of $8x$?				
	(A) x^2	(B) $4x$	(C) $12x$	(D) 16	1
2	$2(x + 5) =$				
	(A) $2x + 5$	(B) $2x + 7$	(C) $2x + 10$	(D) $10x$	1
3	$3(p + q) =$				
	(A) $3p + q$	(B) $3p + 3q$	(C) $3pq$	(D) $p + 3q$	1
4	Which is **not** a factor of $6ab$?				
	(A) ab	(B) $6a$	(C) 4	(D) $3b$	1
5	$5(a - 3) =$				
	(A) $5a - 15$	(B) $5a - 8$	(C) $5a - 3$	(D) $a + 2$	1
6	What is a common factor of $4x$ and 2?				
	(A) 2	(B) 4	(C) x	(D) $4x$	1
7	$4(2x + 7) =$				
	(A) $6x + 11$	(B) $6x + 28$	(C) $8x + 11$	(D) $8x + 28$	1
8	$6(7 - y) =$				
	(A) $13 - y$	(B) $42 - y$	(C) $42 - 6y$	(D) $42 - 7y$	1
9	What is a correct factorisation of $6x - 3$?				
	(A) $3(2x - 1)$	(B) $6(x - 2)$	(C) $3(x - 2)$	(D) $3(2x)$	1
10	$8(3t - 1) =$				
	(A) $16t$	(B) $24t$	(C) $24t - 1$	(D) $24t - 8$	1
11	Which is a common factor of $10a^2$ and $15ab$?				
	(A) 2	(B) $3b$	(C) $5a$	(D) $15a^2$	1
12	$x(x + 5) =$				
	(A) $2x + 5$	(B) $7x$	(C) $x^2 + 5$	(D) $x^2 + 5x$	1
13	What is **not** a common factor of $6x^2$ and $12xy$?				
	(A) $2x^2$	(B) $3x$	(C) 6	(D) x	1
14	$5(2a + 3b - 5c) =$				
	(A) $7a + 8b - 10c$	(B) $10a + 15b - 25c$	(C) $25ab - 10c$	(D) 0	1
15	Which shows the expression $8a - 10ab$ fully factorised?				
	(A) $2(4a - 5ab)$	(B) $4(2a - 5b)$	(C) $a(8 - 10b)$	(D) $2a(4 - 5b)$	1

Expanding and Factorising Calculator Allowed **(Level 1 Test)**

PART B Write answers only in the space provided.

16 Expand $5(a + 6)$. ________________ 1

17 Write 70 as a product of its prime factors. ________________ 1

18 What is the largest number that will divide evenly into $24p$ and $42q$? ________________ 1

19 Expand $7(2x - 1)$. ________________ 1

20 $3(5a + b) =$ ________________ 1

21 Factorise $3x + 30$. ________________ 1

22 Factorise $14a - 12$. ________________ 1

23 Expand $a(a + 4)$. ________________ 1

24 Factorise $x^2 - 9x$. ________________ 1

25 $12(3m - 2n) =$ ________________ 1

PART C Show all working.

26 Expand $-4(3a + 2b) - 5(3c - 2d)$. 1 mark

____

b $7(2n - 3) + 2(3n - 5)$ 2 marks

____

____

c $x(x + y) + y(y + x)$ 2 marks

____

____

d $4(2a + 7) - 3(a + 1)$ 2 marks

____

____

27 Paige and Cooper were asked to factorise $24x + 36$. Paige gave the answer as $4(6x + 9)$ and Cooper said $3(8x + 12)$. Who was correct? Justify your answer. 2 marks

____

____

____

____

28 Expand and simplify:

a $5(3x + 2y) + 4(5x + 2y)$ 2 marks

____

____

29 Factorise fully:

a $9x^2 - 12xy$ 2 marks

____

b $16abc + 8a$ 2 marks

____

Total marks /40

LEVEL 2 TEST PART A

Time allowed: 40 minutes

Question	Marks
1 $9(10e-3) =$ (A) $19e-12$ (B) $90e-12$ (C) $90e-27$ (D) $117e$	1
2 Which is **not** a factor of $36xy^2$? (A) $9xy$ (B) $4x^2$ (C) $6y^2$ (D) $12y$	1
3 $7(3x-4y) =$ (A) $10x-11y$ (B) $21x-11y$ (C) $11x-28y$ (D) $21x-28y$	1
4 $m(m-n) =$ (A) $2m-n$ (B) $2m-mn$ (C) m^2-n (D) m^2-mn	1
5 What is a common factor of $12ab^2$ and $16a^2b$? (A) $4a^2b^2$ (B) $2ab$ (C) $12a^2b$ (D) 8	1
6 $2a(a-3) =$ (A) a^2-3a (B) $2a^2-3a$ (C) a^2-6a (D) $2a^2-6a$	1
7 Which does **not** have a factor of $(x+3)$? (A) $4x+12$ (B) $7x+28$ (C) $xy+3y$ (D) x^2+3x	1
8 $-3(x-4) =$ (A) $-3x+12$ (B) $-3x-12$ (C) $-3x-4$ (D) $-3x+4$	1
9 $3x(2x-5) =$ (A) $5x^2-15x$ (B) $6x^2-15x$ (C) $5x-15$ (D) $6x-15$	1
10 What is the greatest common factor of $9x^2$ and $18xy$? (A) $3x$ (B) $3xy$ (C) $9x$ (D) $18x$	1
11 Which is a correct factorisation of $8ab-12b$? (A) $4a(2b-3)$ (B) $2b(4a-10)$ (C) $4b(2a-3)$ (D) $2(4ab-12b)$	1
12 $-2(7-m) =$ (A) $5-2m$ (B) $5+2m$ (C) $-14-2m$ (D) $2m-14$	1
13 $-(2p+5) =$ (A) $-7p$ (B) $-2p+5$ (C) $-2p-5$ (D) $2p-5$	1
14 $4ab(2a-5b) =$ (A) $8a^2b-20ab^2$ (B) $8ab^2-20a^2b$ (C) $8a^2-20b^2$ (D) $-12a^2b^2$	1
15 Which shows the expression $20x^2-24xy$ fully factorised? (A) $4x(5x-6y)$ (B) $4(5x^2-6xy)$ (C) $2(10x^2-12xy)$ (D) $2x(10x-12y)$	1

PART B Write answers only in the space provided.

16 Expand $9(5x - 4)$. ______________ [1]

17 $a(a - 7) =$ ______________ [1]

18 $10(2x + 3y - 4z) =$ ______________ [1]

19 Factorise $16a + 18b$. ______________ [1]

20 Expand $-2(4x + 3)$. ______________ [1]

21 What is the expansion of $-3(5m - 2n)$? ______________ [1]

22 Factorise fully: $12x^2 - 20xy$. ______________ [1]

23 Complete: $-5x + 35 = -5($______$)$ [1]

24 Expand $-(x - y)$. ______________ [1]

25 Fully factorise $28ab^2 + 14a^2$. ______________ [1]

PART C Show all working.

26 Expand and simplify:

a $-7n - (5 - n)$ [2 marks]

____

____

b $2a(3a + 5) + 3a(a - 4)$ [2 marks]

____

____

c $3(4t - u) - 2(6t + u)$ [2 marks]

____

____

d $7(x^2 - 2) - x(x - 2)$ [2 marks]

____

____

27 Chloe and Kurt were asked to factorise $9b - 3a$. Chloe put $3(3b - a)$ and Kurt put $-3(a - 3b)$. Who was correct? Justify your answer. [1 mark]

____

____

____

28 Find the value of $6(3x + 2y) + 4(2x - 3y)$ when $x = 10$. [2 marks]

____

____

29 Factorise fully:

a $42x^2y - 28xy^2 + 35x^2y^2$ [2 marks]

____

____

b $10a^3b^3c^2 - 25a^2b^3c^3$ [2 marks]

____

____

Total marks ___ /40

LEVEL 3 TEST

PART A

Time allowed: 40 minutes

					Marks
1	$n(n-2) =$				
	(A) $2n-2$	(B) n^2-2n	(C) n^2-2	(D) $2n^2-2n$	1
2	$-a(5a+4) =$				
	(A) $4a-4$	(B) $-5a^2-4a$	(C) $-5a^2+4a$	(D) $-4a^2+4a$	1
3	Which is **not** a factor of $20pqr^2$?				
	(A) $2pq$	(B) $4r^2$	(C) $5pq^2$	(C) $10pqr$	1
4	$5-(x-3) =$				
	(A) $8-x$	(B) $-5x+8$	(C) $2-x$	(D) $-5x+2$	1
5	$-3n(5-2n) =$				
	(A) $-15n-6n^2$	(B) $6n^2-15n$	(C) $2n-16n^2$	(D) $5n^2-8n$	1
6	$2p+3(p+2) =$				
	(A) $5p+6$	(B) $5p+2$	(C) $6p+6$	(D) $6p^2+6$	1
7	Which is **not** a factorisation of $24x^2-12xy$?				
	(A) $4x(6x-3y)$	(B) $3(8x^2-4xy)$	(C) $6x(4x-2y)$	(D) $12(2x^2-y)$	1
8	$10+2(a+4) =$				
	(A) $2a+14$	(B) $12a+8$	(C) $12a+48$	(D) $2a+18$	1
9	Which shows $24x^2y-36xy^2$ fully factorised?				
	(A) $12x(2xy-3y^2)$	(B) $6xy(4x-6y)$	(C) $12xy(2x-3y)$	(D) $24xy(x-3y)$	1
10	$2(3x+4)+4(2x+3) =$				
	(A) $14x+7$	(B) $14x+20$	(C) $11x+13$	(D) $11x+7$	1
11	$-3(2x-1) =$				
	(A) $-6x+3$	(B) $-6x-3$	(C) $-5x+3$	(D) $-5x-3$	1
12	Of which expression is $5x(2x-3y)$ a factorisation?				
	(A) $10x-15y$	(B) $7x^2-15xy$	(C) $10x^2-15xy$	(D) $10x-15xy$	1
13	Which is **not** a correct factorisation of $16t-8tx$?				
	(A) $-8t(x-2)$	(B) $8(2t-tx)$	(C) $-8(-2t+tx)$	(D) $8t(2-tx)$	1
14	$-9a-2a(a-4) =$				
	(A) $-2a^2-a$	(B) $a-2a^2$	(C) $-17a-2a^2$	(D) $8a-11a^2$	1
15	For which expression is $(2x-3)$ **not** a factor?				
	(A) $10x^2-15x$	(B) $18-12x$	(C) $14xy^2-21y$	(D) $6x^2y^2-9xy^2$	1

PART B Write answers only in the space provided.

16 What is the greatest common factor of $6x^2yz$ and $21xyz^2$? ______ [1]

17 Expand $5k(k - l + 7)$. ______ [1]

18 Factorise $9x - 45xy$. ______ [1]

19 Expand $-2a(4a - 7)$. ______ [1]

20 What is the expansion of $-(11 + k)$? ______ [1]

21 Expand $b^2(a + c)$. ______ [1]

22 $-p(3p + 4) =$ ______ [1]

23 Fully factorise $8xy^2 + 4y^2$. ______ [1]

24 $ab(a - b) =$ ______ [1]

25 Complete $8p - 2q =$ ____$(q - 4p)$. [1]

PART C Show all working.

26 **a** Expand and simplify $x(x + 4) + x(x + 5)$. [1 mark]

______

b Factorise the answer to part **a**. [1 mark]

______

c Briefly explain how you could find the answer to part **b** without first expanding and simplifying. [1 mark]

______

______

27 Expand and simplify:

a $5(9n - 4) - 6(4n + 5) + 2(3n - 1)$ [2 marks]

______

______

b $9(m^2 - 2) - m(m - 3) + 5m + 7$ [2 marks]

______

______

c $8(p + 5) - (3p - 1) + 5(7 - p)$ [2 marks]

______

______

28 Factorise fully:

a $12p^3q^2 - 9p^2q^3 - 15pq^2$ [2 marks]

______

______

b $a^3b^3c^3 + a^2b^3c^2 + a^3b^2c^3 + a^2b^2c^2$ [2 marks]

______

______

c $4x^2 + 8xy - 24xy^2 + 4x - 20x^2y + 16x^3$ [2 marks]

______

______

Total marks ___ /40

PRETEST

Time allowed: 15 minutes **Total marks: 15**

					Marks
1	If $x = 5$, what is the value of $2x + 4$?				
	(A) 14	(B) 18	(C) 19	(D) 29	1
2	The sum of 3 and what other number is 12?				
	(A) 4	(B) 8	(C) 9	(D) 15	1
3	$5x + 8 - 2x =$				
	(A) $7x - 8$	(B) $7x + 8$	(C) $3x + 8$	(D) $11x$	1
4	What is the solution to $a + 17 = 35$?				
	(A) $a = 52$	(B) $a = 42$	(C) $a = 18$	(D) $a = 16$	1
5	What is the solution to $m - 29 = 33$?				
	(A) $m = 62$	(B) $m = 52$	(C) $m = 6$	(D) $m = 4$	1
6	What is the solution to $18 - x = 13$?				
	(A) $x = 5$	(B) $x = -5$	(C) $x = 31$	(D) $x = -31$	1
7	If $3x = 105$, then $x =$				
	(A) 315	(B) 108	(C) 102	(D) 35	1
8	What is the solution to $\frac{n}{2} = 18$?				
	(A) $n = 9$	(B) $n = 16$	(C) $n = 20$	(D) $n = 36$	1
9	If $x + y = z$, which is correct?				
	(A) $x = y + z$	(B) $x = z - y$	(C) $x = y - z$	(D) $x = yz$	1
10	If $a = 6$, what is the value of $\frac{2a}{3} + 4$?				
	(A) 8	(B) $5\frac{1}{3}$	(C) $12\frac{2}{3}$	(D) $6\frac{2}{3}$	1
11	For which of these values of x will $2x + 7 = 13$?				
	(A) $x = 2$	(B) $x = 3$	(C) $x = 10$	(D) $x = 12$	1
12	If $m = 7$, what is the value of $\frac{5m - 3}{2}$?				
	(A) 76	(B) 19	(C) 16	(D) 64	1
13	If $12 - 3p = 15$, then $p =$				
	(A) 1	(B) -1	(C) 9	(D) -9	1
14	If $x = \frac{2}{3}$, what is the value of $4x - 1$?				
	(A) $-\frac{1}{3}$	(B) $\frac{1}{3}$	(C) $1\frac{1}{3}$	(D) $1\frac{2}{3}$	1
15	If $y = -1.5$, what is the value of $6y + 7$?				
	(A) 11.5	(B) -2	(C) 2	(D) 16	1

Total marks /15

Equations Calculator Allowed

LEVEL 1 TEST PART A

Time allowed: 40 minutes

					Marks
1	If $2x + 5 = 17$ then $x =$				
	(A) 6	(B) 11	(C) 24	(D) 44	1
2	If $30 - 4x = 10$ then $x =$				
	(A) -10	(B) -5	(C) 5	(D) 10	1
3	What is the solution to $3x - 9 = 9$?				
	(A) $x = 0$	(B) $x = 6$	(C) $x = 18$	(D) $x = 54$	1
4	If $\frac{a}{3} + 4 = 7$, then $a =$				
	(A) 1	(B) 3	(C) 9	(D) 33	1
5	What is the solution to $\frac{n}{2} - 3 = 5$?				
	(A) $n = 1$	(B) $n = 4$	(C) $n = 8$	(D) $n = 16$	1
6	What is the solution to $5x + 24 = 3x + 32$?				
	(A) $x = 1$	(B) $x = 4$	(C) $x = 7$	(D) $x = 28$	1
7	What is the solution to $\frac{x+3}{3} = 3$?				
	(A) $x = 6$	(B) $x = 3$	(C) $x = 1$	(D) $x = 0$	1
8	The solution to $\frac{2k}{7} - 1 = 1$ is:				
	(A) $k = 0$	(B) $k = 1$	(C) $k = 2$	(D) $k = 7$	1
9	If $3x + 6 = -12$, then $x =$				
	(A) 6	(B) -6	(C) 2	(D) -2	1
10	For which of these equations is $x = 2$ a solution?				
	(A) $7x - 5 = 19$	(B) $8x + 12 = 26$	(C) $19 - 3x = 13$	(D) $9x - 26 = 4x$	1
11	If $\frac{2n+7}{3} = 15$ then $n =$				
	(A) 12	(B) 19	(C) 26	(D) 33	1
12	If $6a + 2 = 11$ then $a =$				
	(A) 1.5	(B) $2\frac{1}{6}$	(C) 54	(D) 78	1
13	The solution to $\frac{2m}{3} = 18$ is:				
	(A) $m = 12$	(B) $m = 16$	(C) $m = 17\frac{1}{3}$	(D) $m = 27$	1
14	Which equation does **not** have a solution of $x = 6$?				
	(A) $3x + 5 = x + 17$	(B) $3(x + 2) = 24$	(C) $3x - 6 = 2x$	(D) $3x + 5 = 17 - x$	1
15	If $\frac{a+6}{4} = -2$, then $a =$				
	(A) -14	(B) -12	(C) -4	(D) -2	1

PART B Write answers only in the space provided.

16 What is the solution to $2x + 7 = 15$? ________________ [1]

17 The solution to $5x + 2 = 7 - x$ is $x = 1$. True or false? ________________ [1]

18 If $8y = 5$, what is the value, as a decimal, of y? ________________ [1]

19 What is the solution to $\frac{x}{2} - 3 = 9$? ________________ [1]

20 The equation $\frac{4n + 3}{5} = 11$ has solution $n = 13$. True or false? ________________ [1]

21 What is the solution to $9(x - 2) = 72$? ________________ [1]

22 What is the solution to $29 - 4x = 5$? ________________ [1]

23 If $\frac{x + 4}{3} = 8$, what is the value of x? ________________ [1]

24 What number, when multiplied by 5 and added to 6, gives 11? ________________ [1]

25 What is the solution to $7x - 3 = -3$? ________________ [1]

PART C Show all working.

26 Solve:

a $9x + 5 = 4x + 15$ [2 marks]

b $3(2x - 5) = 17$ [2 marks]

c $\frac{7x + 3}{2} = 208$ [2 marks]

d $12x - 19 = x - 74$ [2 marks]

e $\frac{4x}{3} + 8 = 12$ [2 marks]

f $18 + \frac{x}{3} = 2$ [2 marks]

27 **a** Give two values of x for which $x^2 = 9$. [2 marks]

b Explain why $x^2 = -9$ does not have a solution. [1 mark]

Total marks /40

Equations Calculator Allowed

LEVEL 2 TEST — PART A

Time allowed: 40 minutes

Marks

1 What is the solution to $5x - 3 = 9$?

(A) $x = 30$ (B) $x = 60$ (C) $x = 1.2$ (D) $x = 2.4$ — 1

2 If $5x + 2 = x + 22$, then $x =$

(A) 4 (B) 5 (C) 6 (D) 7 — 1

3 What is the solution to $\frac{n}{5} + 4 = 9$?

(A) $n = 1$ (B) $n = 2.6$ (C) $n = 25$ (D) $n = 65$ — 1

4 What is the solution to $2x - 5 = -9$?

(A) $x = -2$ (B) $x = -7$ (C) $x = 2$ (D) $x = 7$ — 1

5 If $\frac{4m + 7}{3} = 5$, then $m =$

(A) 2 (B) $5\frac{1}{2}$ (C) 6 (D) 9 — 1

6 The solution to $\frac{t}{2} - 5 = -3$ is:

(A) $t = 4$ (B) $t = 1$ (C) $t = -4$ (D) $t = -16$ — 1

7 If $5x - 3 = 21 - x$, then $x =$

(A) 3 (B) 4 (C) 4.5 (D) 6 — 1

8 If $9x + 80 = x + 40$, then $x =$

(A) 4 (B) 5 (C) -4 (D) -5 — 1

9 If $7(x - 3) = 21$, then $x =$

(A) 0 (B) 3 (C) 6 (D) 9 — 1

10 If $\frac{3a}{4} + 5 = 17$, then $a =$

(A) 9 (B) 16 (C) $16\frac{1}{2}$ (D) $29\frac{1}{3}$ — 1

11 If $\frac{2b + 9}{4} = 7$, then $b =$

(A) 9.5 (B) -4 (C) 17.5 (D) 5 — 1

12 Which of these equations has a solution of $x = -4$?

(A) $5x + 3 = -23$ (B) $3x - 2 = -10$ (C) $9x - 1 = -35$ (D) $7x + 17 = -11$ — 1

13 What is the solution of $9q - 3 = 4q + 7$?

(A) $q = 1$ (B) $q = 2$ (C) $q = 3$ (D) $q = 4$ — 1

14 If $\frac{2n - 9}{3} = 5$, then $n =$

(A) 3 (B) 9 (C) 12 (D) 21 — 1

15 What is the solution to $7p + 3 = 3p - 9$?

(A) $p = 1.5$ (B) $p = -1.5$ (C) $p = -3$ (D) $p = 3$ — 1

Equations Calculator Allowed (Level 2 Test)

PART B Write answers only in the space provided.

16 What is the solution to $3x - 4 = 17$? ______________ [1]

17 What is the solution to $8x + 35 = 3$? ______________ [1]

18 A solution of $x^2 = 25$ is $x = -5$. True or false? ______________ [1]

19 If $3(2x - 5) = 21$, what is the value of x? ______________ [1]

20 What is the solution to $\frac{x}{5} + 8 = -2$? ______________ [1]

21 What is the solution to $\frac{x+8}{5} = -2$? ______________ [1]

22 The solution of $7x + 64 = 200 - x$ is $x = 17$. True or false? ______________ [1]

23 What is the solution to $\frac{2x}{3} = 6$? ______________ [1]

24 The solution to $5x + 9 = 3x - 13$ is $x = -2$. True or false? ______________ [1]

25 If $\frac{x}{7} - 3 = 8$, what is the value of x? ______________ [1]

PART C Show all working.

26 Solve:

a $11x - 9 = 4x + 19$ [2 marks]

b $x + 37 = 95 - x$ [2 marks]

c $\frac{3x+2}{5} = 0$ [2 marks]

d $15x + 7 = 42 - 5x$ [2 marks]

e $\frac{3x}{4} + 5 = 9$ [2 marks]

f $6(5x + 7) = 15$ [2 marks]

27 'When 17 is added to 5 times a number and the result multiplied by 4, the answer is 508.'

a Let n be the number. Write an equation to represent the information above. [1 mark]

b Solve the equation to find the number. [2 marks]

Total marks /40

LEVEL 3 TEST — PART A

Time allowed: 40 minutes

Marks

1 If $2p - 5 = -11$, then $p =$

(A) -8 (B) 8 (C) -3 (D) 3 [1]

2 What is the solution to $\frac{d}{4} + 6 = 2$?

(A) $d = -1$ (B) $d = -16$ (C) $d = -32$ (D) $d = 32$ [1]

3 If $\frac{5a}{3} + 4 = 7$, then $a =$

(A) 1.8 (B) 5 (C) 6.5 (D) $18\frac{1}{3}$ [1]

4 What is the solution to $4(x + 3) = 18$?

(A) $x = 1\frac{1}{2}$ (B) $x = 3\frac{3}{4}$ (C) $x = 5\frac{1}{4}$ (D) $x = 7\frac{1}{2}$ [1]

5 Which equation does **not** have a solution of $a = \frac{1}{2}$?

(A) $6a + 12 = 15$ (B) $8a - 4 = 0$ (C) $10a - 7 = -2$ (D) $7a + \frac{1}{2} = 8$ [1]

6 If $7x - 3 = 2x + 8$, then $x =$

(A) $\frac{5}{9}$ (B) 1 (C) $1\frac{2}{9}$ (D) $2\frac{1}{5}$ [1]

7 What is the solution to $6x - 5 = 15 - 2x$?

(A) $x = 1.25$ (B) $x = 1.5$ (C) $x = 2.25$ (D) $x = 2.5$ [1]

8 The equation $x^2 = 13$ has how many solutions?

(A) 0 (B) 1 (C) 2 (D) more than 2 [1]

9 The solution to $12x + 3 = 5x - 4$ is:

(A) $x = -1$ (B) $x = 1$ (C) $x = -\frac{1}{7}$ (D) $x = \frac{1}{7}$ [1]

10 If $5x + 7 - 3x + 2 = 28$ then $x =$

(A) 2.875 (B) 9.5 (C) 11.5 (D) 16.5 [1]

11 The solution to $6(2x - 3) = 15$ is:

(A) $x = 1.5$ (B) $x = 2.75$ (C) $x = 3$ (D) $x = 4.125$ [1]

12 If $\frac{x}{4} - \frac{1}{8} = \frac{2}{3}$, then $x =$

(A) $2\frac{1}{6}$ (B) $2\frac{5}{6}$ (C) $3\frac{1}{6}$ (D) $3\frac{5}{6}$ [1]

13 Which equation has a solution of $x = -1.4$?

(A) $7x + 25 = 13 - 3x$ (B) $11x - 17 = 6x - 24$ (C) $9x - 2 + 6x + 18 = 5$ (D) $6x - 12 = x - 20$ [1]

14 The solution to $\frac{8x - 2}{4} = 6$ is:

(A) $x = 4$ (B) $x = 3\frac{3}{4}$ (C) $x = 3\frac{1}{4}$ (D) $x = 2\frac{3}{4}$ [1]

15 If $\frac{2a}{3} - 4 = -9$, then $a =$

(A) $-7\frac{1}{2}$ (B) $-11\frac{1}{2}$ (C) $-15\frac{1}{2}$ (D) $-19\frac{1}{2}$ [1]

PART B Write answers only in the space provided.

16 What is the solution to $9x - 13 = 50$? ________________ 1

17 What is the solution to $53 - 4x = 17$? ________________ 1

18 If $\frac{x}{19} - 37 = 8$, what is the value of x? ________________ 1

19 What is the solution to $4(3x + 11) = x$? ________________ 1

20 If $\frac{5x + 3}{2} = -8$, what is the value of x? ________________ 1

21 If $x > 0$, what is the solution to $x^2 = 7$, giving the answer to two decimal places? ________________ 1

22 What is the solution to $\frac{3m}{2} + 16 = 49$? ________________ 1

23 What is the solution to $7p + 59 = 17$? ________________ 1

24 The solution to $19x + 52 = 107 + 7x$ is $x = 4\frac{7}{12}$. True or false? ________________ 1

25 What is the solution to $\frac{n}{4} + 39 = -46$? ________________ 1

PART C Show all working.

26 Solve:

a $7p + 18 = -3p + 5$ 2 marks

____

____

____

b $8m + 5 - 2m = m + 17$ 2 marks

____

____

____

c $4(6a - 5) + 7 = 2a$ 2 marks

____

____

____

d $-3(5n + 2) = 16$ 2 marks

____

____

____

e $\frac{23v - 14}{6} = 51$ 2 marks

____

____

____

27 Find the value of x for which $8x + \frac{3}{4} = 2\frac{1}{3}$, expressing the answer as a recurring decimal. 2 marks

____

____

28 Mary and Julie both spent the same amount on pet food at a shop. Mary bought 7 cans of dog food and 3 cans of cat food. Julie bought 4 cans of the dog food and 9 cans of the cat food.

a The cans of cat food cost 85 cents each. Let $C be the cost of each can of dog food. Form an equation from the information. 1 mark

____

____

b Find the cost of a can of dog food. 1 mark

____

____

____

c What amount did Mary spend? 1 mark

____

Total marks /40

Linear Relationships Calculator Allowed

PRETEST

Time allowed: 15 minutes **Total marks: 14**

Marks

Questions 1 to 3 use the number plane on the right.

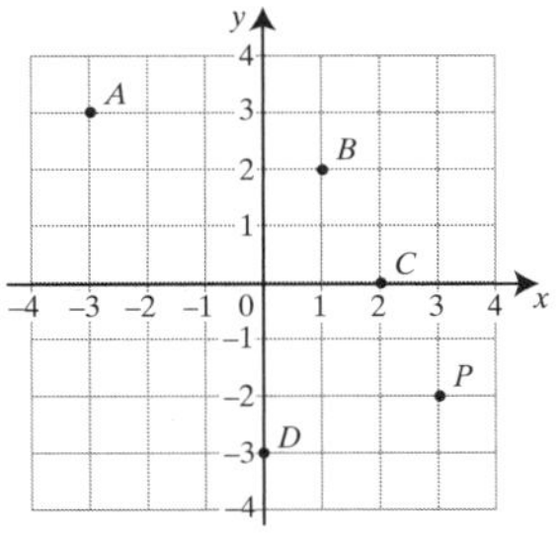

1 What are the coordinates of P?

(A) (2, –3) (B) (3, –2) (C) (–3, 2) (D) (–2, 3) 1

2 Which point lies on the x-axis?

(A) A (B) B (C) C (D) D 1

3 Which point has 0 as its x-value?

(A) A (B) B (C) C (D) D 1

4 What is the next number in this pattern? 4, 11, 18, 25, ?

(A) 31 (B) 32 (C) 33 (D) 34 1

5 What is the missing number in this pattern? 9, 13, ? , 21, 25

(A) 15 (B) 16 (C) 17 (D) 18 1

6 The rule for a pattern is 'begin with 8 and add 9 each time'. What is the fourth number in this pattern?

(A) 35 (B) 37 (C) 42 (D) 44 1

7 What is the next number in this pattern? 87, 81, 75, 69, ?

(A) 65 (B) 64 (C) 63 (D) 61 1

8 What is the missing value? 1112, 1105, ? , 1091, 1084, 1077

(A) 1100 (B) 1099 (C) 1098 (D) 1097 1

Matilda is making a pattern of 'houses' with matches. It is used in questions 9 to 12.

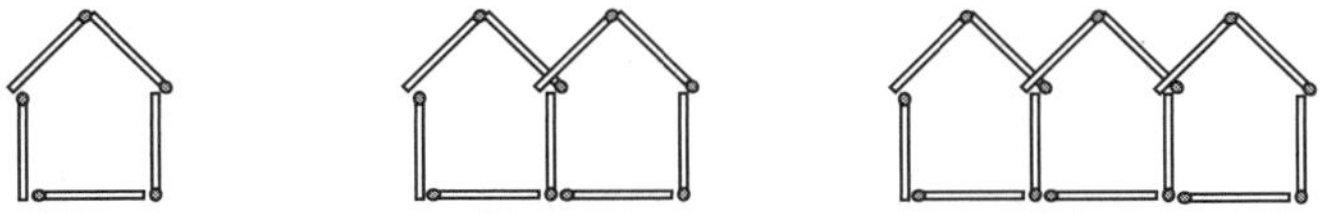

9 How many matches will be needed to make 7 houses?

(A) 17 (B) 19 (C) 27 (D) 29 1

10 How could this rule be completed? 'The number of matches is:

(A) 5 times the number of houses.'
(B) 5 times the number of houses minus 1.'
(C) 4 times the number of houses minus 1.'
(D) 4 times the number of houses plus 1.' 1

11 How many matches would be needed for 101 houses?

(A) 403 (B) 405 (C) 504 (D) 505 1

12 How many houses can be made from 101 matches?

(A) 22 (B) 23 (C) 24 (D) 25 1

13 $4x - 5 + 3x + 2 =$

(A) $7x - 3$ (B) $7x - 7$ (C) $x - 7$ (D) $x - 3$ 1

14 If $x = 6$, then $3x - 5 =$

(A) 13 (B) 3 (C) 31 (D) none of these 1

Total marks 14

Linear Relationships   Calculator Allowed

LEVEL 1 TEST   PART A

Time allowed: 40 minutes

Marks

1 If $m = 5d - 3$, what is the value of m when $d = 17$?

(A) 82   (B) 88   (C) 4   (D) 514   1

2 If $s = 2t + 1$, what is the value of t when $s = 9$?

(A) 19   (B) 17   (C) 4   (D) 5   1

3 What could this be the graph of?

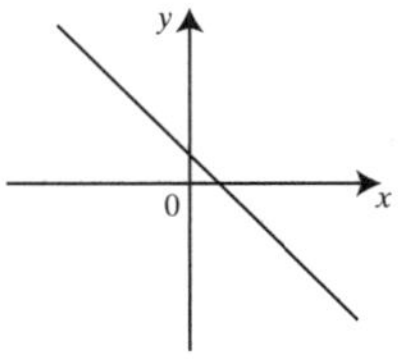

(A) $y = x + 3$ (B) $y = 2x + 2$ (C) $y = 2 - x$ (D) $y = 3x - 1$ 1

4 What is the missing value in this table of values?

x	0	1	2	3	4
y	7	10	13		19

(A) 14   (B) 16   (C) 17   (D) 18   1

5 What is the rule for the table of values in question 4?

(A) $y = x + 7$ (B) $y = 2x + 7$ (C) $y = 3x + 7$ (D) $y = 4x + 7$ 1

6 What is the point of intersection of these two lines?

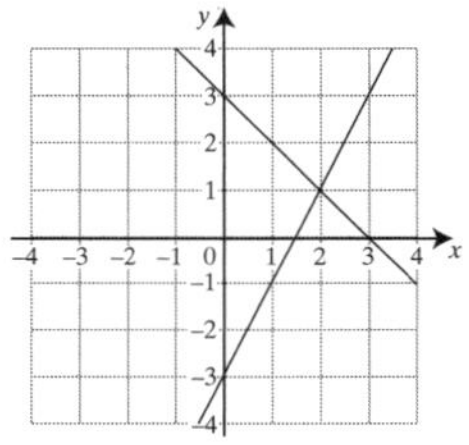

(A) (1, 2)   (B) (2, 1)   (C) (3, 0)   (D) (0, 3)   1

7 Which of these points will the line $y = 4x - 1$ pass through?

(A) (0, 1)   (B) (1, 2)   (C) (2, 7)   (D) (5, 3)   1

8 Which table of values is correct for the rule $y = 7x + 3$?   1

(A)

x	1	2	3
y	10	13	16

(B)

x	1	2	3
y	10	17	20

(C)

x	1	2	3
y	7	10	13

(D)

x	1	2	3
y	10	17	24

9 The graph shows the line $y = 2x - 5$. For what value of x will $y = 19$?

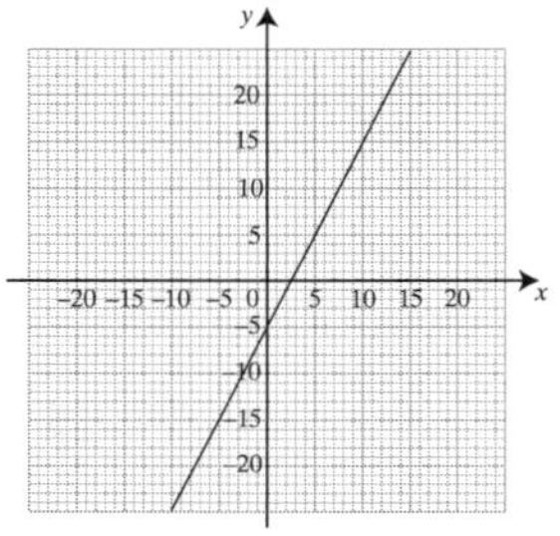

(A) 9   (B) 12   (C) 13   (D) 33   1

10 Referring to the graph, for what value of x does $2x - 5 = -7$?

(A) -6 (B) -1 (C) 1   (D) 6   1

11 Which of these points will **not** lie on the line $y = 3 - 2x$?

(A) (0, 3)   (B) (1, 1)   (C) (−1, 5)   (D) (2, −2)   1

This pattern of shapes is made with matches. It is used for questions 12 to 14.

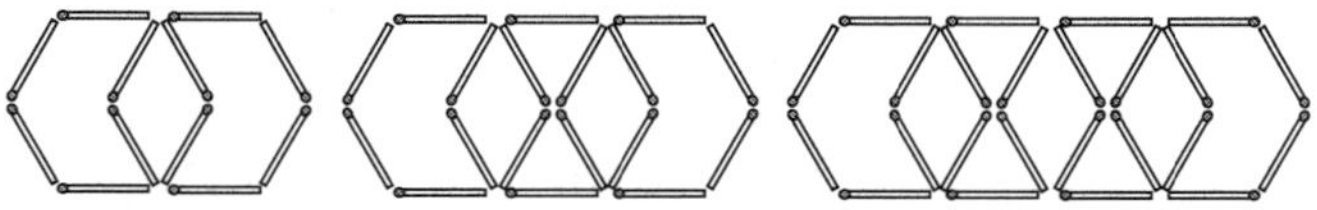

Shape 1 12 matches   Shape 2 18 matches   Shape 3 24 matches

12 How many matches will be needed for Shape 5?

(A) 30   (B) 32   (C) 34   (D) 36   1

13 What rule could be used to link the number of matches, m, with the shape number, s?

(A) $m = s + 6$ (B) $m = 2s + 6$ (C) $m = 6s$ (D) $m = 6(s + 1)$ 1

14 What shape number will need 60 matches?

(A) 9   (B) 10   (C) 11   (D) 12   1

15 Which shows the graph of $y = 2x - 3$?

(A) 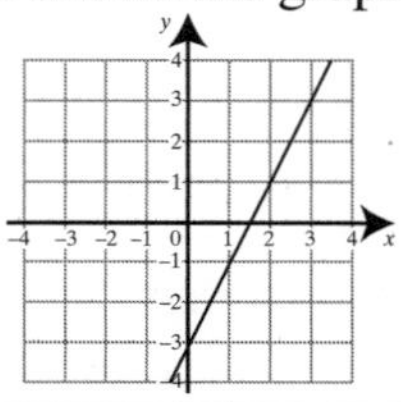

(B) 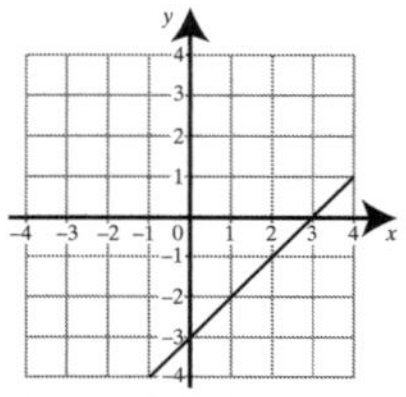

(C) 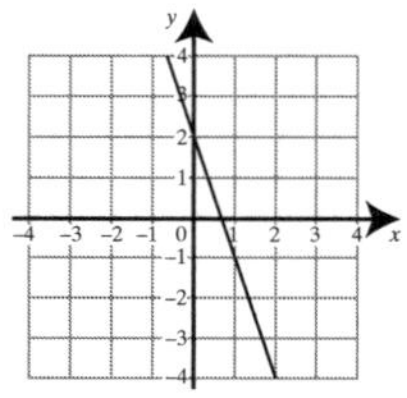

(D) 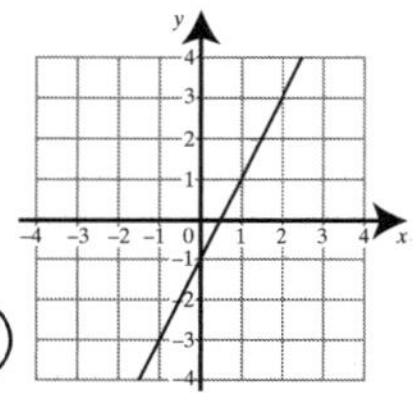

1

PART B Write answers only in the space provided.

16 If $x = 8$, what is the value of $2x - 7$? ________________ 1

17 If $m = 5p + 4$, for what value of p does $m = 24$? ________________ 1

18 Given that $y = 9 - 4x$, what is the value of y when $x = 3$? ________________ 1

19 If $y = 4x + 3$ for what value of x will $y = 3$? ________________ 1

20 Fill in the missing value in this table. 1

x	0	1	2	3	4	5
y	8	15		29	36	43

21 Write a rule for the table of values in question 20. ________________ 1

22 The line $y = 3x + ♦$ will pass through the point (2, 10).

What number does ♦ represent? ________________ 1

Use this graph to answer questions 23 to 25.

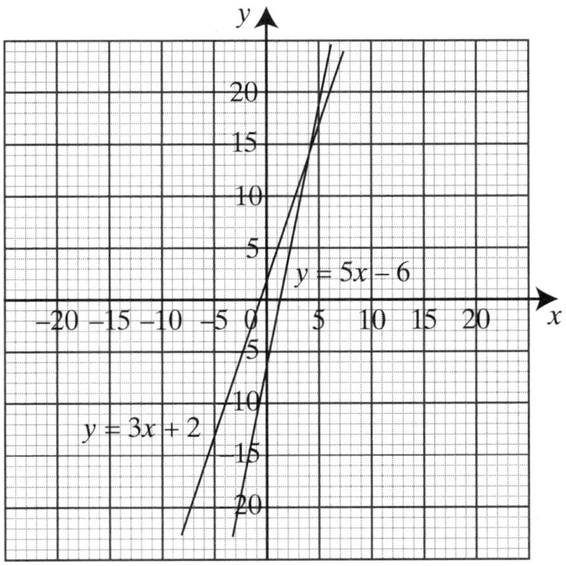

23 What is the point of intersection of the lines $y = 3x + 2$ and $y = 5x - 6$? ________________ 1

24 What is the solution to the equation $3x + 2 = 20$? ________________ 1

25 What is the solution to the equation $5x - 6 = -1$? ________________ 1

PART C Show all working.

26 This is a pattern of hexagons made with matches.

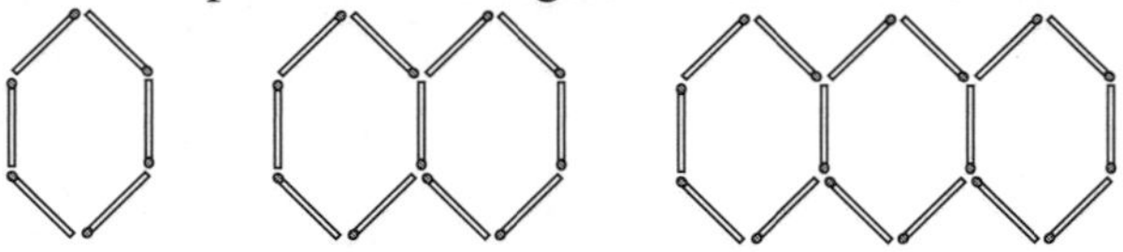

a Complete the table of values for the pattern. [2 marks]

Number of hexagons (h)	1	2	3	4	5	6
Number of matches (m)	6	11				

b Complete in words: The number of matches is equal to [1 mark]

____

____

c Write the rule as an equation using numbers and symbols. [1 mark]

____

d How many matches would be needed if there were 50 hexagons? [1 mark]

____

e How many hexagons would there be if 301 matches were used? [1 mark]

____

27 a Complete the table of values for $y = 2x + 3$. [2 marks]

x	0	1	2	3
y				

b Plot the points on the number plane and draw the line $y = 2x + 3$. [2 marks]

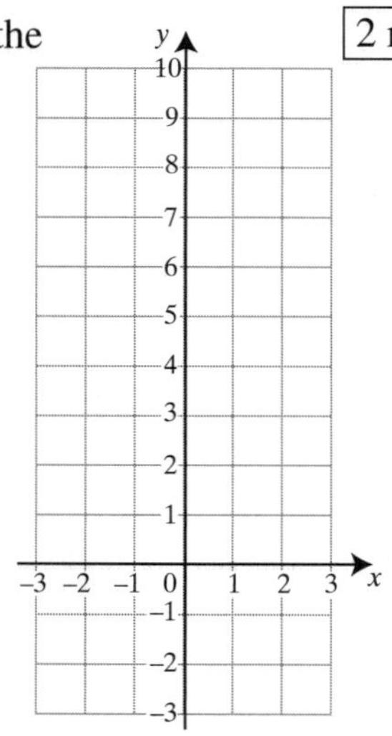

c Use the graph to solve the equation $2x + 3 = -1$. [1 mark]

____

28 The graph shows the lines $y = x + 1$, $y = 2x + 1$, $y = -x + 1$ and $y = -2x + 1$.

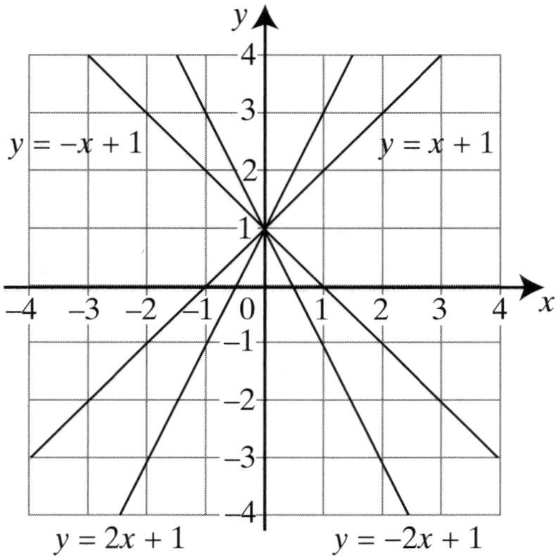

a What feature on the graph do all the lines have in common? [1 mark]

____

____

b What feature do all the equations have in common? [1 mark]

____

____

c What is the effect of the constant term in an equation of a line? [1 mark]

____

____

d What is the effect of a negative coefficient of x? [1 mark]

____

____

Total marks /40

LEVEL 2 TEST — PART A

Time allowed: 40 minutes

Marks

1 If $y = 7x - 3$, what is the value of y when $x = 5$?

Ⓐ 31 Ⓑ 32 Ⓒ 33 Ⓓ 34 — 1

2 Which of these points lies on the line $y = 6x + 5$?

Ⓐ (–1, 1) Ⓑ (–3, –23) Ⓒ (–2, –7) Ⓓ (–4, 19) — 1

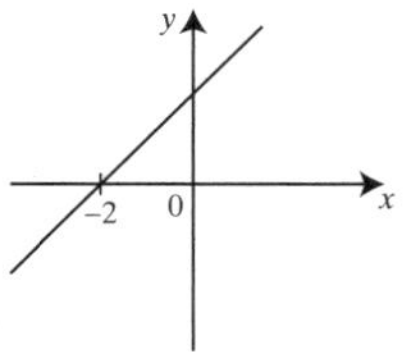

3 Referring to the graph on the right, what could be the equation of this line?

Ⓐ $y = x - 2$ Ⓑ $y = 2 - x$ Ⓒ $y = -2x$ Ⓓ $y = x + 2$ — 1

4 At what point could the line in the graph on the right cut the y-axis?

Ⓐ (2, 0) Ⓑ (0, 2) Ⓒ (–2, 0) Ⓓ (0, –2) — 1

5 What is the missing value in this table of values?

x	1	2	3	5	11
y	7	11	15	23	

Ⓐ 27 Ⓑ 31 Ⓒ 35 Ⓓ 47 — 1

The graph on the right shows the lines $y = 2x + 5$ and $y = -x - 1$. It is used in questions 6, 7 and 8.

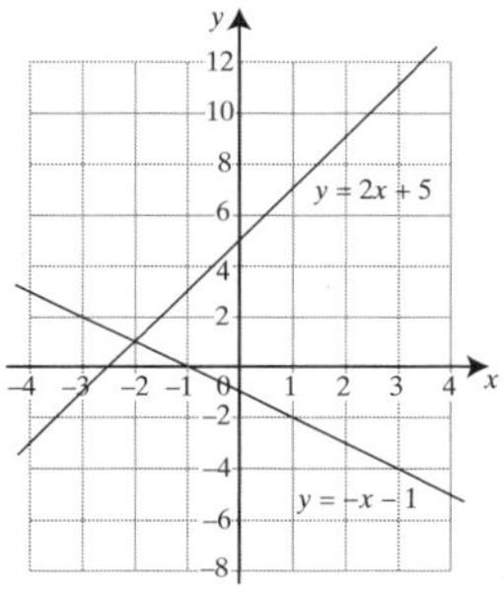

6 What is the point of intersection of the two lines?

Ⓐ (1, –2) Ⓑ (–1, 2) Ⓒ (–2, 1) Ⓓ (2, –1) — 1

7 What is the solution to the equation $2x + 5 = 13$?

Ⓐ $x = 2$ Ⓑ $x = 3$ Ⓒ $x = 4$ Ⓓ $x = 8$ — 1

8 Which is the same line as $y = -x - 1$?

Ⓐ $y = x + 1$ Ⓑ $y = -(x - 1)$ Ⓒ $y = -(x + 1)$ Ⓓ $y = 1 - x$ — 1

9 What equation represents the points in the table of values?

x	1	2	3	4
y	6	10	14	18

Ⓐ $y = x + 5$ Ⓑ $y = 2x + 4$ Ⓒ $y = 3x + 3$ Ⓓ $y = 4x + 2$ — 1

10 What will be the value of y when $x = 28$ if $y = 3x - 7$?

Ⓐ 77 Ⓑ 73 Ⓒ 91 Ⓓ 321 — 1

11 Which point will the line $y = 7 - 2x$ **not** pass through?

Ⓐ (1, 5) Ⓑ (2, 3) Ⓒ (5, –3) Ⓓ (–4, 1) — 1

12 If $p = 3m + 5$, for what value of m will $p = 62$?

Ⓐ 19 Ⓑ 21 Ⓒ 22 Ⓓ 23 — 1

13 Which is **not** a linear relationship?

Ⓐ $y = 2x$ Ⓑ $y = x^2$ Ⓒ $y = 7$ Ⓓ $y = 4 - x$ — 1

14 'To get the number of matches, multiply the number of squares by 3 and take away 4.' If m is the number of matches and s the number of squares, the rule could be written as:

Ⓐ $m = 3s - 4$ Ⓑ $m = 4 - 3s$ Ⓒ $s = 3m - 4$ Ⓓ $s = 4 - 3m$ — 1

15 Which is the graph of $y = 4 - 2x$?

Ⓐ 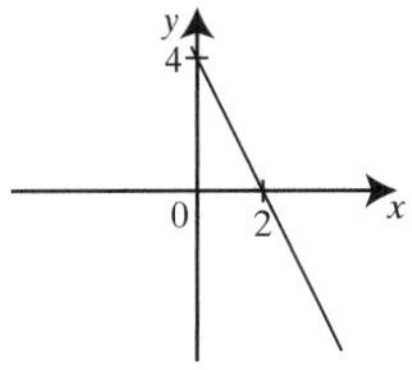

Ⓑ 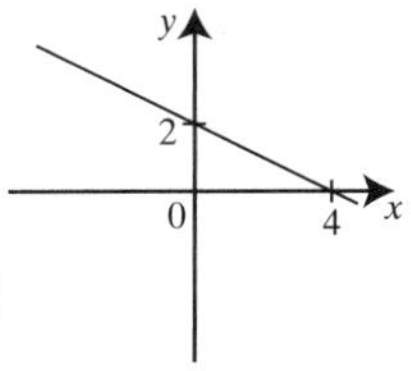

Ⓒ 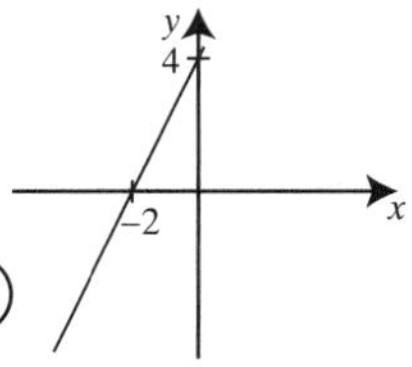

Ⓓ 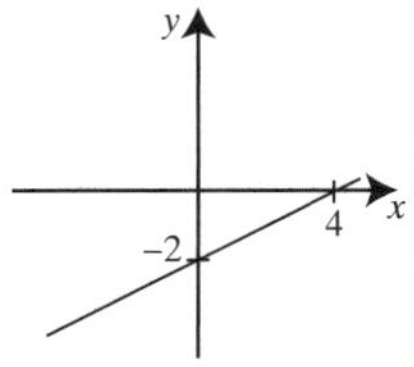

1

PART B Write answers only in the space provided.

16 If $y = 8x - 5$, what is y when $x = 12$? ________________ 1

17 Fill in the missing value for $y = 5x - 4$. 1

x	0	1	2
y	–4	1	

18 What is the rule for the relationship shown in this table of values? ________________ 1

x	0	1	2	3
y	2	8	14	20

19 At what point does the line $y = 2x + 5$ intercept the y-axis? ________________ 1

20 Write in symbols: 'To find the value of y multiply the value of x by 7 and then add 2.' ____________ 1

21 Will the line $y = 3x - 1$ increase or decrease? ________________ 1

22 The point $(-2, -3)$ lies on the line $y = 2x - 1$. True or false? ________________ 1

23 If $m = 5t + 2$ for what value of t will $m = 102$? ________________ 1

24 The line $y = 1 - x$ will pass through the origin. True or false? ________________ 1

25 What is the coefficient of x in the equation $y = 6x + 4$? ________________ 1

PART C Show all working.

26 Consider this pattern:

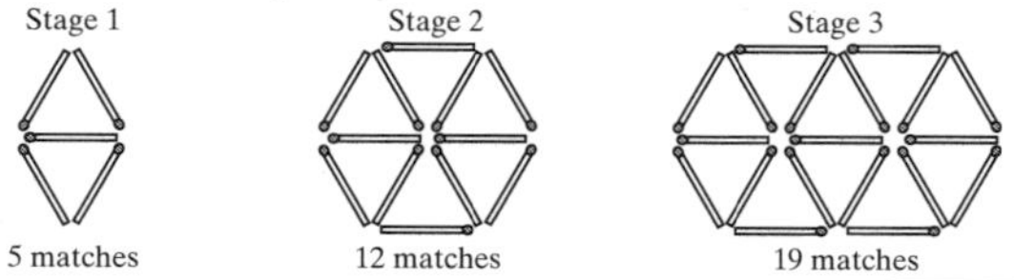

a Fill in the table of values for this pattern, using s for the stage number and m for the number of matches. 2 marks

s	1	2	3	4	5	6	7
m	5	12	19				

b Write a rule in words to explain how the number of matches is linked to the stage number. 1 mark

____

____

c Write the rule using numbers and symbols. 1 mark

____

d How many matches would be needed at Stage 15? 1 mark

____

e At what stage would 180 matches be needed? 1 mark

____

27 a Complete the table of values for $y = 2x - 7$. 1 mark

x	0	1	2
y			

b Complete this table of values for $y = 5 - x$. 1 mark

x	0	1	2
y			

c Graph the lines $y = 2x - 7$ and $y = 5 - x$. 2 marks

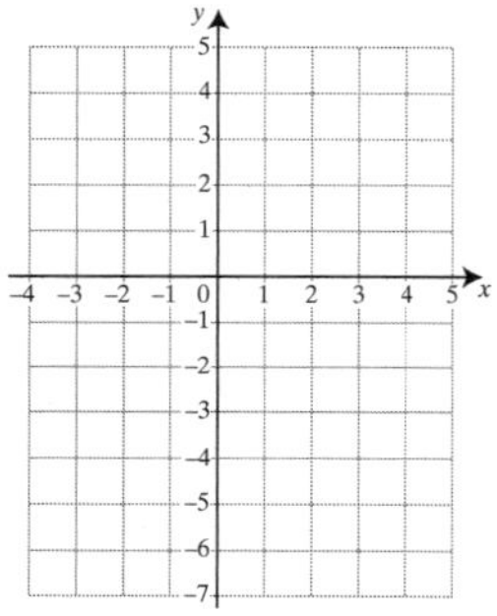

d What is the point of intersection of the two lines? 1 mark

____

e For what value of x will $2x - 7 = 5$? 1 mark

____

28 The graph shows the lines $y = 4x$, $y = 4x - 4$ and $y = 4x + 6$.

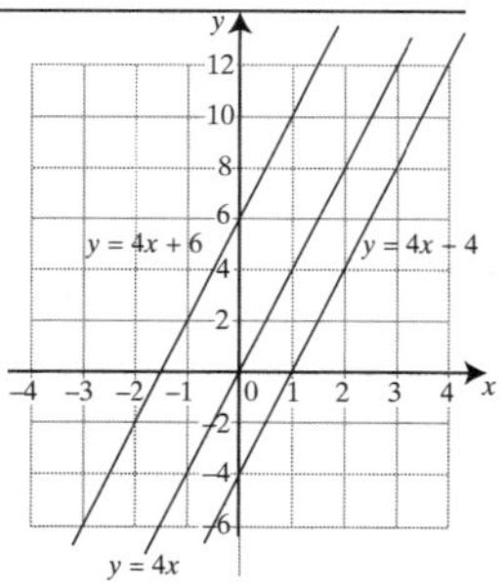

a What similarities are in the equations of the lines and what feature does this create in the graphs? 2 marks

____

____

b What is the significance of the constant term in the equation? 1 mark

____

Total marks /40

Linear Relationships Calculator Allowed

LEVEL 3 TEST — PART A

Time allowed: 40 minutes

Marks

1 What is the coefficient of x in the equation $y = 5x - 2$?

(A) 5 (B) $5x$ (C) 2 (D) -2 — 1

2 If $y = -2x + 7$, what is the value of y when $x = 4$?

(A) 1 (B) -1 (C) -15 (D) -17 — 1

3 Which of these lines will **not** pass through the point $(-2, 3)$?

(A) $y = 3x + 9$ (B) $y = -3x - 3$ (C) $y = 2x + 1$ (D) $y = -2x - 1$ — 1

4 Which line will have the greatest slope (gradient)?

(A) $y = x + 7$ (B) $y = 2x + 5$ (C) $y = 3x - 4$ (D) $y = 4x + 1$ — 1

5 If $d = 4p + 5$, what is the value of d when $p = -3$?

(A) -17 (B) -7 (C) 4 (D) 6 — 1

6 Which of these lines will be parallel to $y = 5 - 3x$?

(A) $y = -3x + 4$ (B) $y = -5x - 1$ (C) $y = 3x + 7$ (D) $y = 5x + 2$ — 1

7 Which line will intersect $y = 3x - 6$ on the y-axis?

(A) $y = 6 - 3x$ (B) $y = 4x - 8$ (C) $y = 5x - 6$ (D) $y = 3x + 4$ — 1

8 What is the gradient (slope) of the line $y = 8 - x$?

(A) 8 (B) 1 (C) -1 (D) 0 — 1

9 If $x = 2t - 1$, for what value of t will $x = -9$?

(A) -4 (B) -5 (C) -17 (D) -19 — 1

10 Which is the equation of a different line to the others?

(A) $x + y = 5$ (B) $y = -x + 5$ (C) $y = x - 5$ (D) $y = 5 - x$ — 1

11 Which of these points will **not** lie on the same line as the others?

(A) $(4, 7)$ (B) $(2, 5)$ (C) $(-3, -5)$ (D) $(8, 17)$ — 1

12 What is the equation of the line that has y-intercept -2 and passes through the point $(3, 4)$?

(A) $y = -2x + 10$ (B) $y = 2x - 2$ (C) $y = -2x - 2$ (D) $y = 4x - 8$ — 1

13 What is the point of intersection of $y = 4x - 6$ and $y = -2x + 18$?

(A) $(-2, -14)$ (B) $(-1, 20)$ (C) $(3, 6)$ (D) $(4, 10)$ — 1

14 What is the equation of the line that passes through $(2, 7)$ and $(5, -2)$?

(A) $y = 2x + 3$ (B) $y = 13 - 3x$ (C) $y = 2x - 12$ (D) $y = -4x + 15$ — 1

15 Which of these lines will **not** pass through $(6, 2)$?

(A)
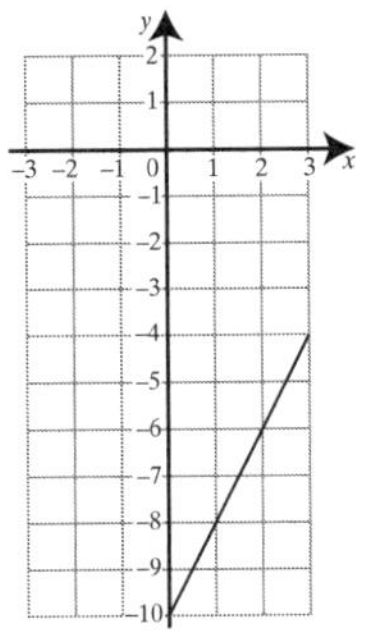

(B)
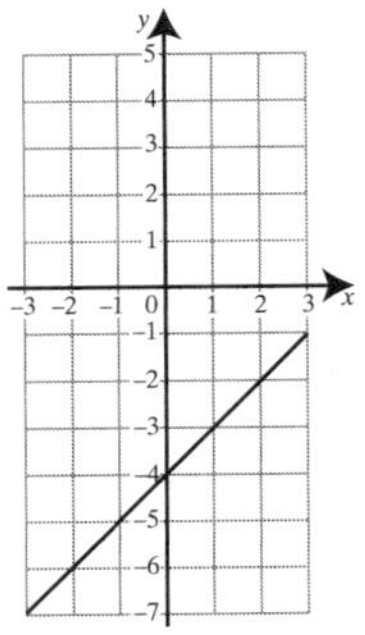

(C)
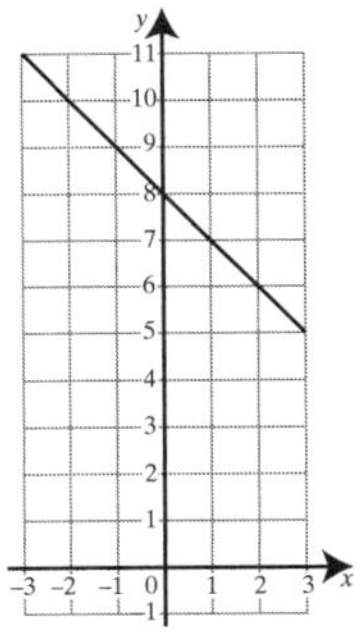

(D)
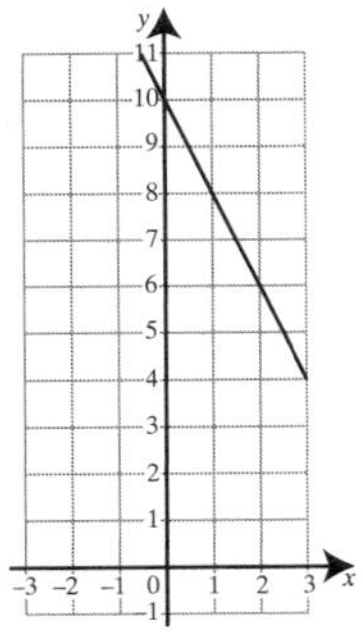

— 1

PART B Write answers only in the space provided.

16 If $y = \frac{1}{2}x + 6$, what is y when $x = 8$? ______________ 1

17 The point (3, 7) lies on the line $y = 13 - 2x$. True or false? ______________ 1

18 What is the missing value in the table of values? ______________ 1

x	1	2	3	5	8
y	53	46	39		4

19 At what point will the line $y = 6x - 2$ intercept the y-axis? ______________ 1

20 If $p = 3m + 1$, what is the value of m for which $p = 52$? ______________ 1

21 What is the coefficient of x in the equation $y = 15 - 3x$? ______________ 1

22 The line $y = 7 + 5x$ will decrease. True or false? ______________ 1

23 If $y = 4x + 3$, what is the value of y when $x = \frac{1}{2}$? ______________ 1

24 What is the constant term in the equation $y = 4x - 9$? ______________ 1

25 Write an equation for the points in this table of values. ______________ 1

x	1	2	3	4	5
y	3	11	19	27	35

PART C Show all working.

26 Callie is making a pattern with pins and string.

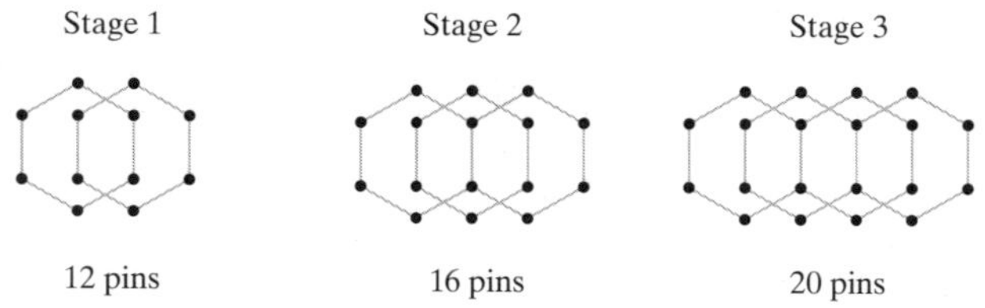

a Complete the table of values. 1 mark

Stage (s)	1	2	3	4	5
Number of pins (p)	12	16	20		

b Write the rule in symbols. 1 mark

____

c How many pins will be needed at Stage 12? 1 mark

____

d What is the first stage that will need more than 100 pins? 1 mark

____

PART C continued Show all working.

27 a Complete the table of values for $y = 4x - 3$. [1 mark]

x	0	1	2
y			

b Graph the line $y = 4x - 3$. [1 mark]

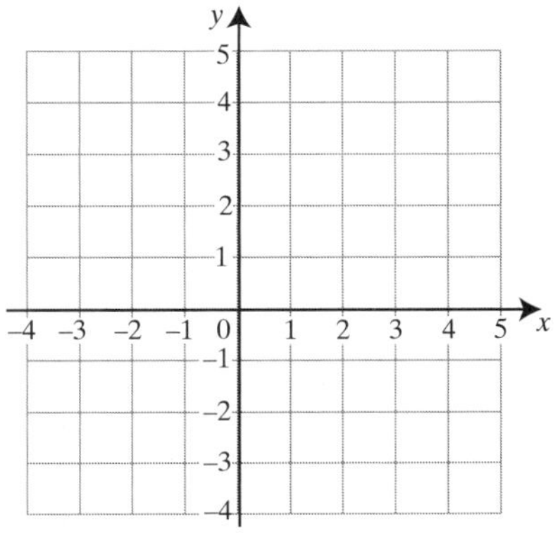

c How many points lie on the line $y = 4x - 3$? [1 mark]

d Do the lines $y = 4x - 3$ and $y = 3 + 4x$ have a point of intersection? Explain why or why not. [1 mark]

28 a What is the gradient (slope) of the line $y = 3x + 2$? [1 mark]

b Briefly explain what that means. [1 mark]

c What does it mean if a gradient is negative? [1 mark]

29 The diagram shows the graph of $y = x^2$.

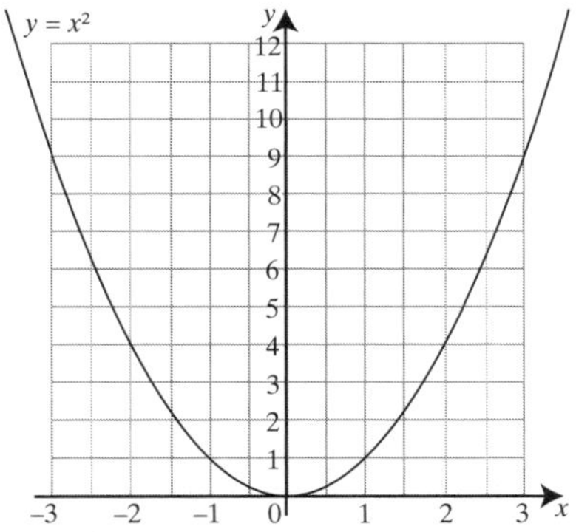

a Is $y = x^2$ a linear relationship? Explain why or why not. [1 mark]

b A line is then drawn on the graph. [1 mark]

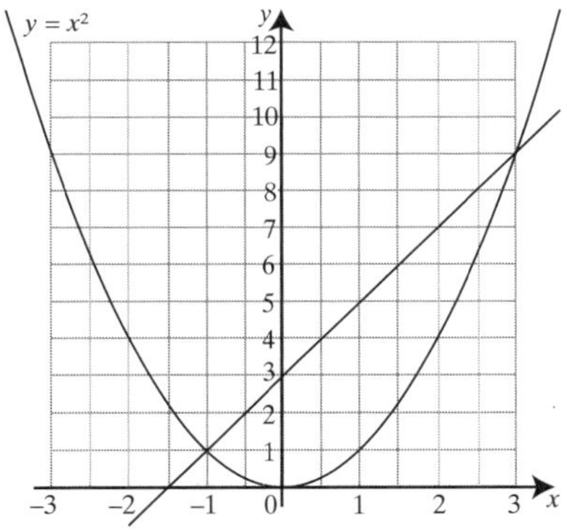

What is the y-intercept of that line?

c What are the coordinates of the points of intersection of the straight line and $y = x^2$? [1 mark]

d What is the equation of the straight line? [1 mark]

Total marks /40

Mixed Algebra and Equations Calculator Allowed

LEVEL 1 TEST PART A

Time allowed: 40 minutes

Marks

1 $5a + 3a =$

(A) $8a$ (B) $8a^2$ (C) $15a$ (D) $15a^2$ [1]

2 $8x \div 4x =$

(A) 2 (B) $2x$ (C) 4 (D) $4x$ [1]

3 If $x = 3$, what is the value of $2x - 1$?

(A) 4 (B) 5 (C) 8 (D) 22 [1]

4 $9p - 3p =$

(A) 6 (B) 3 (C) $6p$ (D) $3p$ [1]

5 $6x \times 5x =$

(A) $11x$ (B) $11x^2$ (C) $30x$ (D) $30x^2$ [1]

6 $2(x + 5) =$

(A) $2x + 7$ (B) $2x + 10$ (C) $x^2 + 7$ (D) $x^2 + 10$ [1]

7 Which is **not** a factor of $6xy$?

(A) $2x$ (B) $3y$ (C) xy (D) 12 [1]

8 $4x + 9x^2 =$

(A) $13x$ (B) $13x^2$ (C) $13x^3$ (D) cannot be simplified [1]

9 What is the missing value in the table?

x	0	1	2	3	4
y	5	9	13		21

(A) 16 (B) 17 (C) 18 (D) 19 [1]

10 Which equation has the solution $x = 9$?

(A) $x + 6 = 3$ (B) $\frac{x}{3} = 3$ (C) $\frac{x+1}{2} = 20$ (D) $3x = 3$ [1]

11 Which is the correct factorisation of $7e - 14$?

(A) $7(e - 14)$ (B) $7(e - 7)$ (C) $7(e - 2)$ (D) cannot be factorised [1]

12 $8x + 7y - 3x + 4y =$

(A) $5x + 11y$ (B) $5x + 3y$ (C) $11x + 11y$ (D) $11x + 3y$ [1]

13 $4(2a + 3b) =$

(A) $20ab$ (B) $24ab$ (C) $6a + 7b$ (D) $8a + 12b$ [1]

14 Which point does **not** lie on the line $y = x$?

(A) (3, 3) (B) (–2, –2) (C) (5, 0) (D) (27, 27) [1]

15 What is the solution to the equation $2x + 3 = 5$?

(A) $x = 1$ (B) $x = 2$ (C) $x = 3$ (D) $x = 4$ [1]

Mixed Algebra and Equations Calculator Allowed **(Level 1 Test)**

PART B Write answers only in the space provided.

16 Expand $3(5x - 2)$. ____________ [1]

17 $6x + 3x + 2x + 4y =$ ____________ [1]

18 Factorise $3x - 15$. ____________ [1]

19 $7a \times 2 =$ ____________ [1]

20 What is the solution to $x - 7 = 7$? ____________ [1]

21 Expand $x(x + y)$. ____________ [1]

22 Factorise $a^2 + 2a$. ____________ [1]

23 If $m = 3$ and $n = 4$, what is the value of mn? ____________ [1]

24 $20pq \div 2p =$ ____________ [1]

25 $-2(4x + 5) =$ ____________ [1]

PART C Show all working.

26 Simplify:

a $2x + 3x \times 4x$ [1 mark]

____

b $\frac{12ab}{6b}$ [1 mark]

____

27 Expand and simplify:

a $x(x + 3) + 3(x + 2)$ [2 marks]

____

____

b $5(3a - 2) + 4(2a + 1)$ [2 marks]

____

____

28 Factorise fully:

a $5x^2 - 10x$ [1 mark]

____

b $8a + 4$ [1 mark]

____

29 Solve:

a $\frac{2x}{3} + 5 = 9$ [2 marks]

____

____

____

b $3(2x - 10) = 0$ [2 marks]

____

____

____

30 **a** Fill in the table of values for the rule $y = 8 - 2x$. [1 mark]

x	0	1	2	3
y				

b Graph the line $y = 8 - 2x$ on the axes provided. [1 mark]

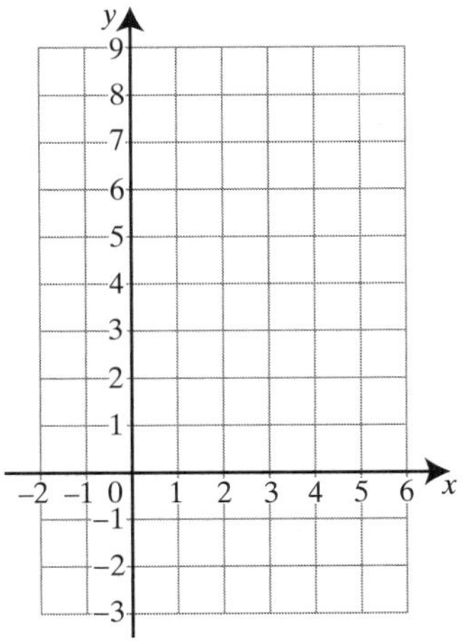

c For what value of x does $8 - 2x = -2$? [1 mark]

____

Total marks /40

Mixed Algebra and Equations Calculator Allowed

LEVEL 2 TEST PART A

Time allowed: 40 minutes

Marks

1 $3x^2 + 5x^2 =$

(A) $8x^2$ (B) $34x^2$ (C) $8x^4$ (D) $34x^4$ 1

2 If $p = -2$, what is the value of $5 - 2p$?

(A) 1 (B) 9 (C) –6 (D) –17 1

3 What is the solution to $2x + 1 = 13$?

(A) $x = 6$ (B) $x = 7$ (C) $x = 24$ (D) $x = 28$ 1

4 $5(2a + 3b) =$

(A) $7a + 8b$ (B) $7a + 15b$ (C) $10a + 8b$ (D) $10a + 15b$ 1

5 $-2x \times 3x =$

(A) $-6x$ (B) $-6x^2$ (C) $-5x$ (D) $-5x^2$ 1

6 What is the greatest common factor of $12a^2b$ and $15ab^2$?

(A) $60a^2b^2$ (B) $60ab$ (C) $3a^2b^2$ (D) $3ab$ 1

7 What is the rule for the values in this table?

x	0	1	2	3	4
y	2	5	8	11	14

(A) $y = x + 2$ (B) $y = 2x + 2$ (C) $y = 3x + 2$ (D) $y = 4x + 2$ 1

8 $3p + 5q - 2p + 4q =$

(A) $p + 9q$ (B) $p + q$ (C) $5p + 9q$ (D) $5p + q$ 1

9 $x(x - 4) =$

(A) $2x - 4$ (B) $x^2 - 4$ (C) $x^2 - 4x$ (D) $-6x$ 1

10 Which is a factor of $9x - 12$?

(A) 9 (B) $x - 3$ (C) $x - 4$ (D) $3x - 4$ 1

11 If $x = 3$ and $y = 2$, what is the value of xy^2?

(A) 36 (B) 12 (C) 18 (D) 1024 1

12 Which equation does **not** have the solution $x = 3$?

(A) $x + 4 = 7$ (B) $4x = 12$ (C) $10 - x = 7$ (D) $x - 9 = 6$ 1

13 Which is **not** correct?

(A) $x^2 - 8x = 2x(x - 4)$ (B) $16x + 20 = 4(4x + 5)$

(C) $ab - 3a = a(b - 3)$ (D) $10 - 5a = 5(2 - a)$ 1

14 Which of these points will lie on the line $y = 4x - 1$?

(A) (1, 2) (B) (2, 7) (C) (3, 9) (D) (4, 17) 1

15 $-3(x + 2) =$

(A) $-3x + 2$ (B) $-3x - 1$ (C) $-3x - 6$ (D) $-3x + 6$ 1

Mixed Algebra and Equations Calculator Allowed (Level 2 Test)

PART B Write answers only in the space provided.

16 What is the solution to $\frac{x}{2} = 10$? ________________ 1

17 Expand $2x(x - 4)$. ________________ 1

18 Factorise fully $8a - 4$. ________________ 1

19 If $a = 5$ and $b = 3$, what is the value of $2b - a$? ________________ 1

20 $6k \div 2k =$ ________________ 1

21 $3x^2 + 2x + x^2 =$ ________________ 1

22 What is the next value in this pattern? 1025, 1013, 1001, 989, ? ________________ 1

23 $3m \times 4m =$ ________________ 1

24 $-2a - a =$ ________________ 1

25 Complete: $-6a - 12 = -6($_______$)$. 1

PART C Show all working.

26 Solve:

a $5x + 8 = 3x - 12$ 2 marks

____

____

____

b $\frac{2x - 5}{3} = 9$ 2 marks

____

____

____

27 Expand and simplify:

a $2x(x + 4) - 7(x + 1)$ 2 marks

____

____

b $3p(p + 5) - 2p(p - 6)$ 2 marks

____

____

28 Factorise $8a^2b + 12ab^2 + 20ab$ fully. 2 marks

____

____

29 a Complete the table of values for $y = x + 3$. 1 mark

x	0	1	2	3
y				

b Complete the table of values for $y = -x - 1$. 1 mark

x	0	1	2	3
y				

c Graph the lines $y = x + 3$ and $y = -x - 1$ on the axes below. 2 marks

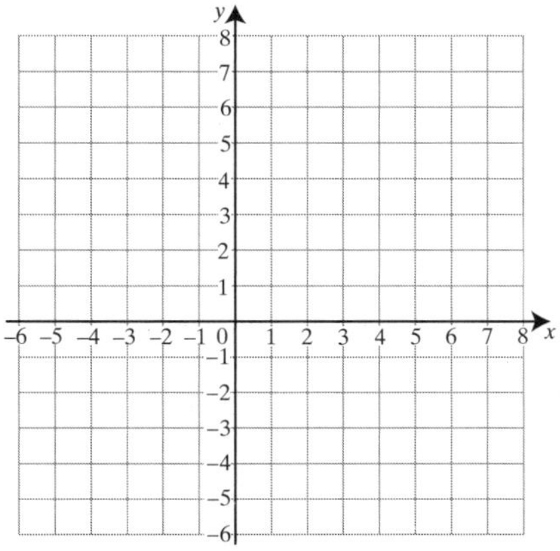

d Write down the solution of the equation $x + 3 = -x - 1$. 1 mark

____

Total marks /40

Mixed Algebra and Equations Calculator Allowed

LEVEL 3 TEST PART A

Time allowed: 40 minutes

Marks

1 $5 - 3(x - 4) =$

(A) $17 - 3x$ (B) $-3x - 7$ (C) $-3x + 7$ (D) $3x + 17$ [1]

2 $16x^2 \div 2x \times 4x =$

(A) 2 (B) $2x$ (C) $32x$ (D) $32x^2$ [1]

3 What is the solution to the equation $\frac{2x}{3} - 4 = 5$?

(A) $x = 1.5$ (B) $x = 5.5$ (C) $x = 9.5$ (D) $x = 13.5$ [1]

4 Which expression is factorised fully?

(A) $12x^2 - 36x = 12(x^2 - 3x)$ (B) $18a^2b - 4ab = ab(18a - 4)$

(C) $20m^2 - 15mn = 5m(4m - 3n)$ (D) $8p^2q - 12pq^2 = 4pq^2(2p - 3)$ [1]

5 Which point does **not** lie on the line $y = 3x - 2$?

(A) $(-3, -7)$ (B) $(0, -2)$ (C) $(5, 13)$ (D) $(37, 109)$ [1]

6 If $a = -4$, what is the value of $2a^2$?

(A) -32 (B) 32 (C) -64 (D) 64 [1]

7 $\frac{24a^2bc^2}{8ab^2c} =$

(A) $\frac{3ac}{b}$ (B) $\frac{3c}{b}$ (C) $\frac{3a}{bc}$ (D) $\frac{3c}{ab}$ [1]

8 $-5m^2 + 2m - 4m^2 + 3m =$

(A) $5m - m^2$ (B) $5m - 9m^2$ (C) $-m^2 - m$ (D) $-9m^2 - m$ [1]

9 How many different factors does $10xy$ have?

(A) 4 (B) 12 (C) 15 (D) 16 [1]

10 Which is **not** the equation of a linear relationship?

(A) $y = 3$ (B) $y = 2x$ (C) $y = \frac{x}{2}$ (D) $y = x^2$ [1]

11 Which expansion is **not** correct?

(A) $-x(x + 5) = -x^2 - 5x$ (B) $-n(7 - n) = n^2 - 7n$

(C) $-4(-3 + t) = -12 - 4t$ (D) $ab(a + b) = a^2b + b^2a$ [1]

12 Which is a correct factorisation of $-8f - 2g + h$?

(A) $-(8f - 2g + h)$ (B) $-(8f + 2g - h)$ (C) $-(8f - 2g - h)$ (D) $-(8f + 2g + h)$ [1]

13 $-2x^2 + 5x \times -3x =$

(A) $-17x^2$ (B) $17x^2$ (C) $-13x^2$ (D) $13x^2$ [1]

14 What is the equation of the line that passes through the points $(-1, -4)$, $(1, 2)$ and $(2, 5)$?

(A) $y = x + 3$ (B) $y = 3x - 1$ (C) $y = 4x - 2$ (D) $y = 5x - 3$ [1]

15 What is the solution to $5x - 9 = 14 - x$?

(A) $x = \frac{5}{6}$ (B) $x = 1\frac{1}{4}$ (C) $x = 3\frac{5}{6}$ (D) $x = 5\frac{3}{4}$ [1]

PART B Write answers only in the space provided.

16 Factorise $3a^2 + 6a + 3$. ______ 1

17 $-5p^2 - 6p + 7p^2 - 2q - 3p^2 + 8p =$ ______ 1

18 $-2a(3a - 5bc) =$ ______ 1

19 What is the solution to the equation $3p - 17 = -2$? ______ 1

20 What is the rule for the values in this table? ______ 1

x	1	2	4	7
y	–4	3	17	38

21 $(4xy \times 3x) \div (2x^2 \times 6y) =$ ______ 1

22 Expand $abc(ab - bc)$. ______ 1

23 At what point will the line $y = 2x - 3$ cut the y-axis? ______ 1

24 If $p = -2$, $q = -3$ and $r = -5$, what is the value of $q - pr$? ______ 1

25 Factorise fully: $-6mn - 3m$. ______ 1

PART C Show all working.

26 Expand and simplify:

a $5pq(2p + 3q) + 4p(3pq - 5)$ 2 marks

______

______

b $4x(2x - 5y) - 3y(6x - 7y) - 2xy(4 + y)$ 2 marks

______

______

27 Solve:

a $3(2x - 5) = 2(5x - 1)$ 2 marks

______

______

______

b $\frac{7x}{4} + 11 = -9$ 2 marks

______

______

______

28 Matilda and Hugo sat for the same test. 2 marks
They were asked to factorise $24xyz^2 + 36x^2yz$. Matilda put $6xyz(6x + 4z)$. Hugo put $4xyz(6z + 9x)$. Neither received full marks. Explain why not and give the answer that would get full marks.

______

______

29 The graph shows the lines $y = 2x - 8$ and $y = 4 - x$.

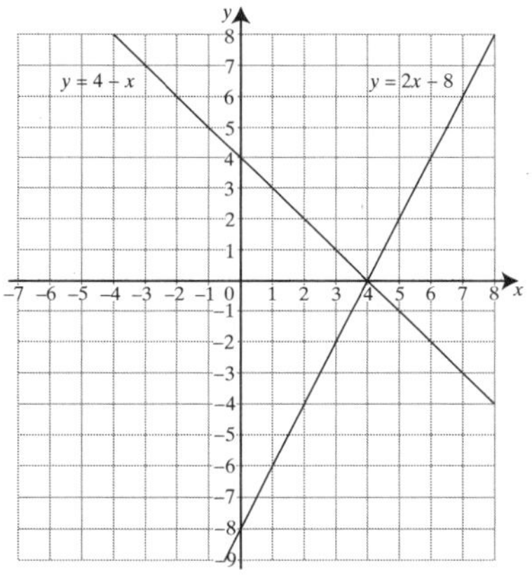

a What is the point of intersection of the lines? 1 mark

______

b Briefly explain the significance of a point of intersection. 1 mark

______

c On the same diagram show the line $y = 2x + 4$. 1 mark

d Explain any similarities between the line $y = 2x + 4$ and the other lines. 2 marks

______

______

______

Total marks /40

Perimeter and Area Calculator Allowed

PRETEST

Time allowed: 15 minutes **Total marks: 16**

1 900 cm =

Ⓐ 0.09 m Ⓑ 90 m Ⓒ 0.9 m Ⓓ 9 m 1

2 How many millimetres are in 80 m?

Ⓐ 8000 Ⓑ 80 000 Ⓒ 800 000 Ⓓ 8 000 000 1

3 Which unit of area is the largest?

Ⓐ 1 m^2 Ⓑ 1 cm^2 C 1 km^2 Ⓓ 1 ha 1

4 What is the shaded area in square units?

Ⓐ 6 Ⓑ 12 Ⓒ 14 Ⓓ 18 1

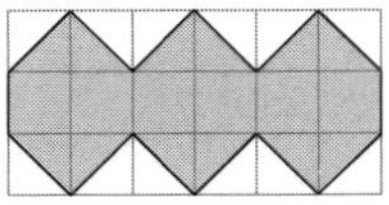

5 What is the area of a square of side 9 cm?

Ⓐ 36 cm^2 Ⓑ 54 cm^2 Ⓒ 72 cm^2 Ⓓ 81 cm^2 1

6 What is the perimeter of a rectangle 8 m long and 5 m wide?

Ⓐ 13 m Ⓑ 18 m Ⓒ 26 m Ⓓ 40 m 1

7 17.6 ha is how many square metres?

Ⓐ 176 000 Ⓑ 17 600 Ⓒ 17 600 000 Ⓓ 1 760 000 1

8 What is the area of a triangle with base length 8 m and perpendicular height 6 m?

Ⓐ 10 m^2 Ⓑ 12 m^2 Ⓒ 24 m^2 Ⓓ 48 m^2 1

9 What is the perimeter of a square of side 5 cm?

Ⓐ 20 cm Ⓑ 25 cm Ⓒ 30 cm Ⓓ 35 cm 1

10 What is the area of a rectangle 1.2 m long and 70 cm wide?

Ⓐ 0.84 m^2 Ⓑ 8.4 cm^2 C 840 cm^2 Ⓓ 8.4 m^2 1

11 What is the area of this triangle?

Ⓐ 132 m^2 Ⓑ 1830 m^2 Ⓒ 330 m^2 Ⓓ 335.5 m^2 1

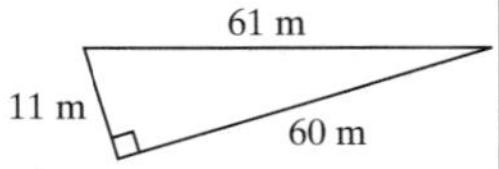

12 A rectangular paddock is 450 m long and 225 m wide. The area of the paddock is closest to:

Ⓐ 1 ha Ⓑ 10 ha Ⓒ 100 ha Ⓓ 1000 ha 1

13 How many metres is 5 m and 8 cm?

Ⓐ 5.8 Ⓑ 5.08 Ⓒ 5.008 Ⓓ none of these 1

14 What is the area of this triangle in square centimetres?

Ⓐ 1125 Ⓑ 1912.5 Ⓒ 540 Ⓓ 900 1

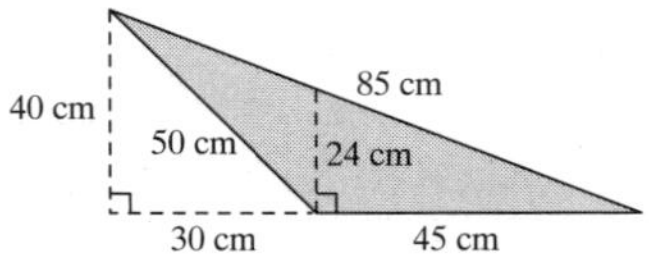

15 6.25 km =

Ⓐ 6025 m Ⓑ 602.5 m Ⓒ 625 m Ⓓ 6250 m 1

16 What is the area of a square with perimeter 24 cm?

Ⓐ 36 cm^2 Ⓑ 32 cm^2 Ⓒ 18 cm^2 Ⓓ 16 cm^2 1

Total marks /16

LEVEL 1 TEST PART A

Time allowed: 40 minutes

1 What is the perimeter of a rhombus that has one side of length 7 cm?

(A) 28 cm (B) 42 cm (C) 49 cm (D) not enough information 1

2 What is the area of this parallelogram?

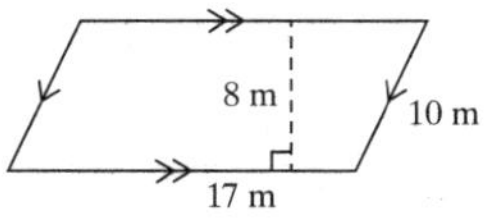

(A) 80 m^2 (B) 136 m^2 (C) 170 m^2 (D) 1360 m^2 1

3 What is the perimeter of the parallelogram in question 2?

(A) 27 m (B) 35 m (C) 54 m (D) 70 m 1

4 Four hundred square millimetres is how many square centimetres?

(A) 20 (B) 40 (C) 2 (D) 4 1

5 What is the perimeter of this shape?

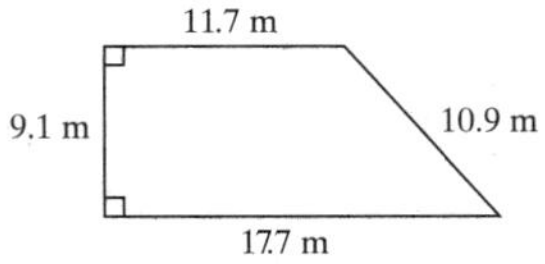

(A) 47.6 m (B) 49.4 m (C) 58.5 m (D) 98.8 m 1

6 What is the area of the shape in question 5?

(A) 127.53 m^2 (B) 160.23 m^2 (C) 161.07 m^2 (D) 133.77 m^2 1

7 One diagonal of a kite is 12 cm long and the other is 5 cm long. What is the area of the kite?

(A) 30 cm^2 (B) 34 cm^2 (C) 60 cm^2 (D) 68 cm^2 1

8 What is the area of this trapezium?

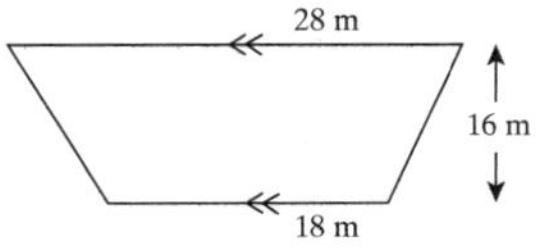

(A) 252 m^2 (B) 288 m^2 (C) 368 m^2 (D) 448 m^2 1

9 What is the area of a rhombus that has one diagonal of length 52 cm and one of 34 cm?

(A) 172 cm^2 (B) 442 cm^2 (C) 884 cm^2 (D) 1768 cm^2 1

10 What is the area of this shape?

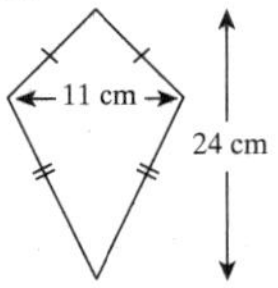

(A) 66 cm^2 (B) 132 cm^2 (C) 264 cm^2 (D) 306.25 cm^2 1

11 What is the area of this shape?

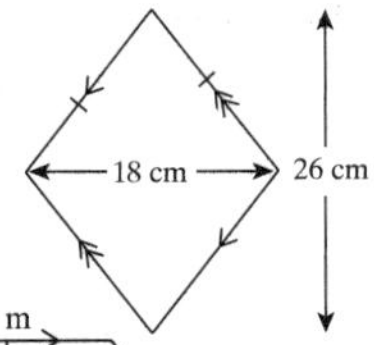

(A) 117 cm^2 (B) 468 cm^2 (C) 484 cm^2 (D) 234 cm^2 1

12 What is the perimeter of this shape?

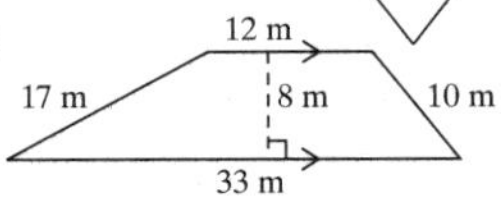

(A) 63 m (B) 70 m (C) 72 m (D) 80 m 1

13 What is the area of the shape in question 12?

(A) 225 m^2 (B) 162 m^2 (C) 382.5 m^2 (D) 180 m^2 1

14 The perimeter of a rhombus is 1.2 m. How many centimetres long is each side?

(A) 20 (B) 30 (C) 40 (D) 60 1

15 *ABCD* is a rectangle. $AB = 3.8$ m, $BC = 1.1$ m and $PQ = 1.3$ m.
What is the area of *ABQP*?

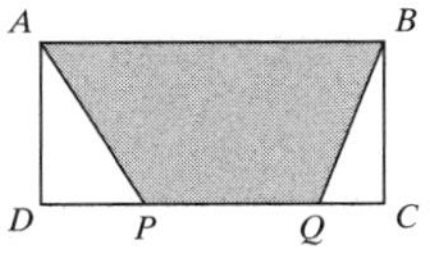

(A) 2.805 m^2 (B) 3.185 m^2 (C) 2.09 m^2 (D) 2.47 m^2 1

PART B Write answers only in the space provided.

16 How many square metres are in 4 ha? ________________ [1]

17 What is the area of a parallelogram with base length 12 cm and perpendicular height 8 cm? ________________ [1]

18 A kite has diagonals of length 9 cm and 7 cm. What is its area? ________________ [1]

19 How many square metres are in 9 km^2? ________________ [1]

20 What is the perimeter of a rhombus of side length 19.5 m? ________________ [1]

21 What is the perimeter in metres of the trapezium on the right? ________________ [1]

22 What is the area of the trapezium in square centimetres? ________________ [1]

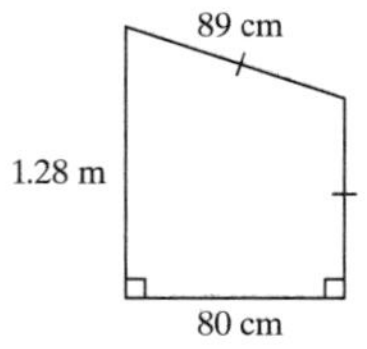

23 A rhombus has its longer diagonal 15 cm long and the shorter one is 12 cm. What is the area? ________________ [1]

24 What is the perimeter, in metres, of this kite? ________________ [1]

25 What is the area of the kite in square metres? ________________ [1]

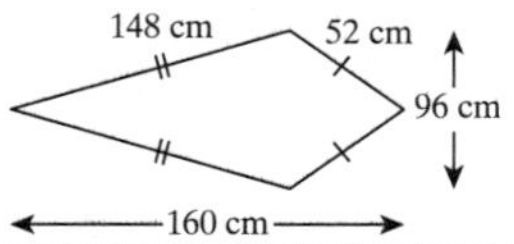

PART C Show all working.

26 The diagram shows a plan of a paddock.

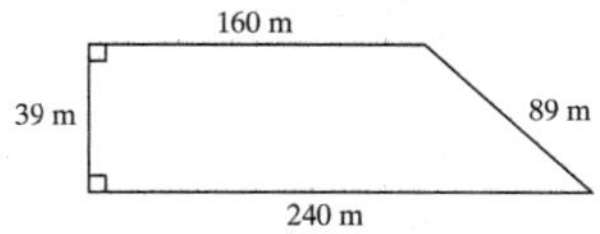

a Find the cost of fencing the paddock at $13 per metre. [2 marks]

b Find the cost of seed for the paddock at $150 per hectare. [2 marks]

27 a What is the area of this shape in square centimetres? [2 marks]

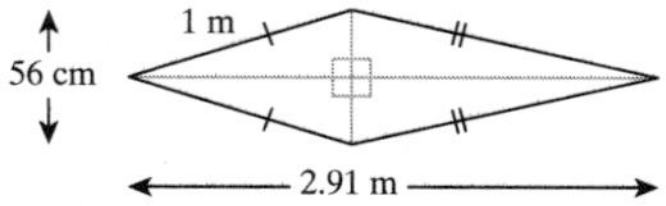

b The perimeter is 5.94 m. What is the length of each of the unknown sides? [2 marks]

28 The diagram shows a frame around a square picture.

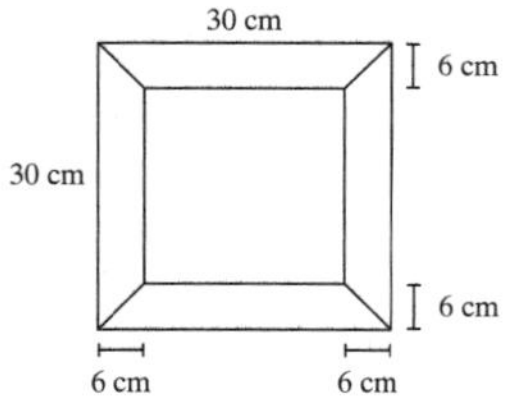

a Briefly explain two different methods that could be used to get the area of the frame. [2 marks]

b What is the area of the frame? [1 mark]

29 The diagram shows a rhombus of side length 8.5 cm.

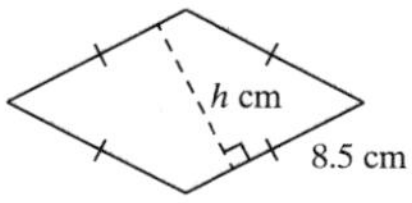

a What is the perimeter of the rhombus? [1 mark]

b One diagonal is 15 cm long and the other is 8 cm long. What is the area of the rhombus? [1 mark]

c What is the value of h? (Give the answer to one decimal place.) [2 marks]

Total marks /40

Perimeter and Area Calculator Allowed

LEVEL 2 TEST PART A

Time allowed: 40 minutes

1 The area of this block of land is closest to:

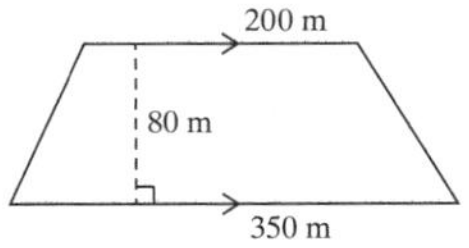

Ⓐ 0.2 ha Ⓑ 2 ha Ⓒ 20 ha Ⓓ 200 ha | 1

2 A parallelogram has base length 1.2 m and perpendicular height 80 cm. What is its area?

Ⓐ 96 cm^2 Ⓑ 0.96 m^2 Ⓒ 9.6 m^2 Ⓓ 96 m^2 | 1

3 What is the perimeter of this shape?

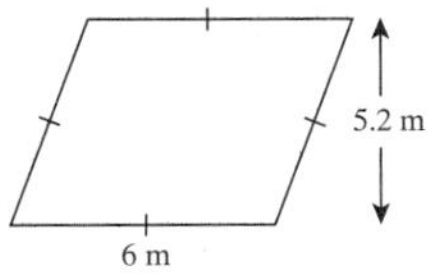

Ⓐ 22.4 m Ⓑ 24 m Ⓒ 20.8 m Ⓓ 36 m | 1

4 What is the area of the shape in question 3?

Ⓐ 15.6 m^2 Ⓑ 18 m^2 Ⓒ 31.2 m^2 Ⓓ 36 m^2 | 1

5 How many square millimetres are in 35 m^2?

Ⓐ 35 000 Ⓑ 350 000 Ⓒ 3 500 000 Ⓓ 35 000 000 | 1

6 What is the area of this shape?

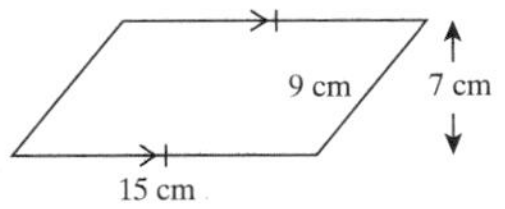

Ⓐ 63 cm^2 Ⓑ 676.5 cm^2 Ⓒ 105 cm^2 Ⓓ 135 cm^2 | 1

7 What is the perimeter of the shape in question 6?

Ⓐ 31 cm Ⓑ 32 cm Ⓒ 44 cm Ⓓ 48 cm | 1

8 How many square centimetres is one square kilometre?

Ⓐ 1 000 000 Ⓑ 10 000 000 Ⓒ 100 000 000 Ⓓ 10 000 000 000 | 1

9 What is the area of this trapezium?

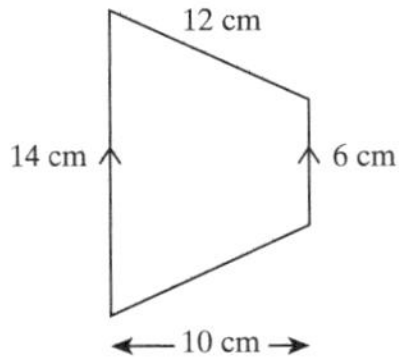

Ⓐ 154 cm^2 Ⓑ 112 cm^2 Ⓒ 100 cm^2 Ⓓ 66 cm^2 | 1

10 A rhombus of side 5 cm has one diagonal 4.8 cm long and the other 8.8 cm long. What is its perimeter?

Ⓐ 20 cm Ⓑ 25 cm Ⓒ 18.6 cm Ⓓ 27.2 cm | 1

11 What is the area of the rhombus in question 10?

Ⓐ 21.12 cm^2 Ⓑ 25 cm^2 Ⓒ 34 cm^2 Ⓓ 42.24 cm^2 | 1

12 What is the perimeter of this kite?

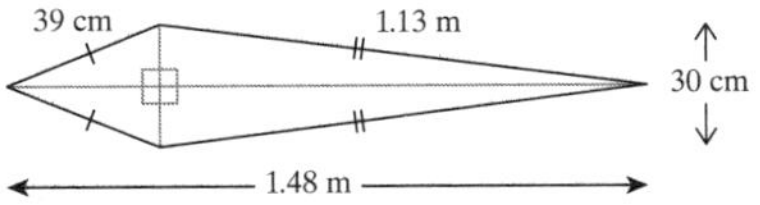

Ⓐ 3.56 m Ⓑ 3.3 m Ⓒ 3.93 m Ⓓ 3.04 m | 1

13 What is the area of the kite in question 12?

Ⓐ 0.222 m^2 B 2203.5 cm^2 Ⓒ 3390 cm^2 Ⓓ 0.2886 m^2 | 1

14 What is the perimeter of this shape?

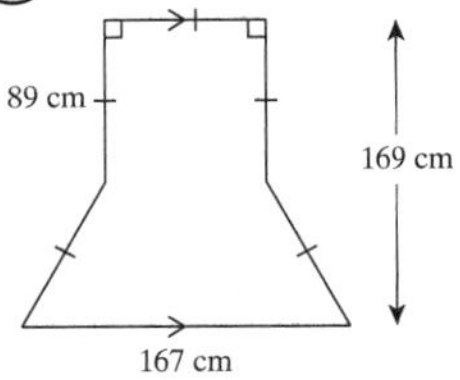

Ⓐ 612 cm Ⓑ 614 cm Ⓒ 781 cm Ⓓ 425 cm | 1

15 What is the area of the shape in question 14?

Ⓐ 10 240 cm^2 Ⓑ 18 161 cm^2 Ⓒ 28 223 cm^2 Ⓓ 29 553 cm^2 | 1

PART B Write answers only in the space provided.

16 How many square centimetres are in 20 m²? ____________ 1

17 What is the area of a square with perimeter 64 cm? ____________ 1

18 What is the perimeter, in metres, of a rhombus of side 78 cm? ____________ 1

19 What is the area, in square metres, of a kite with one diagonal 1.06 m and the other 92 cm? ____________ 1

20 AB = 14.2 cm, AD = 16.9 cm, EF = 12 cm and DC = 38 cm. What is the area of the trapezium? ____________ 1

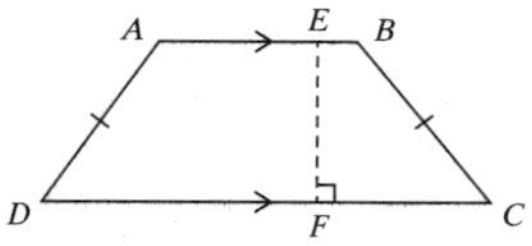

21 What is the perimeter of the trapezium in question 20? ____________ 1

22 $PS = RS$ = 20 cm. PR = 32 cm. The perimeter of the kite is 170 cm. What is the length of PQ? ____________ 1

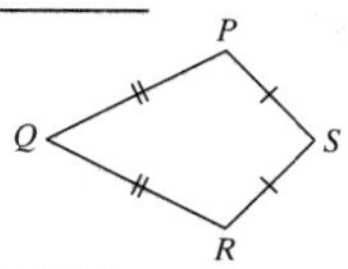

23 The area of the kite in question 22 is 1200 cm². What is the length of SQ? ____________ 1

24 Nine square kilometres is how many hectares? ____________ 1

25 A diagonal of a square is 24 cm long. What is the area of the square? ____________ 1

PART C Show all working.

26 The diagram shows a block of land.

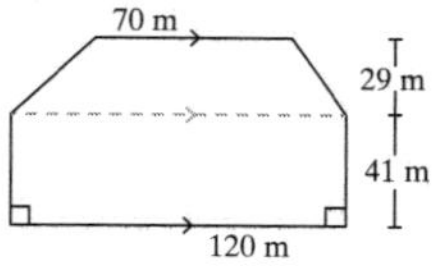

a What is the area of the land? 2 marks

____

b Together with an adjoining block, the total area is 1.5 ha. How many square metres smaller is the adjoining block of land? 2 marks

____

27 This is a diagram of a park.

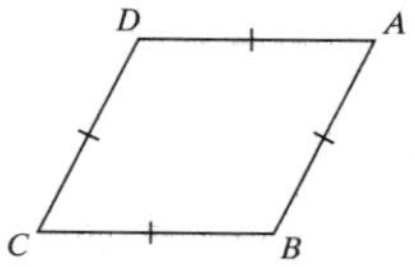

a Ruby walks around the outside edge of the park. She travels 584 m. How long is each side of the park? 1 mark

____

b Cheryl walks from point A to point B then to point C and straight back to point A, which is a distance of 512 m. How long is diagonal AC? 1 mark

____

c Darcy walks from point D to point B and then to point C and back to D. The distance he travels is 100 m less than that of Ruby. How long is diagonal BD? 1 mark

____

d What is the area of the park in square metres? 1 mark

____

28 The diagram shows a kite in a right-angled triangle.

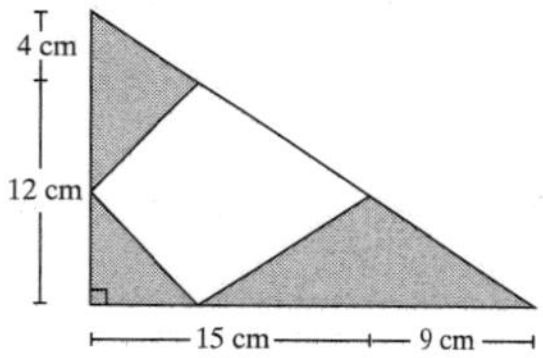

What is the shaded area? 2 marks

____

29 This shape is made up of a rectangle, a parallelogram and a rhombus. The diagonals of the rhombus are 14.4 cm and 10.8 cm.

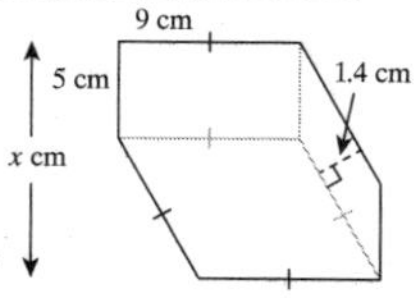

a What is the total area of the shape? 3 marks

____

b What is the value of x? 2 marks

____

Total marks /40

LEVEL 3 TEST PART A

Time allowed: 40 minutes

1 How many square metres is 128 cm^2?

Ⓐ 1.28 Ⓑ 0.128 Ⓒ 0.0128 Ⓓ 0.001 28 1

2 A rhombus has side length of 12 cm and perpendicular height of 9.5 cm. What is its area?

Ⓐ 57 cm^2 Ⓑ 114 cm^2 Ⓒ 144 cm^2 Ⓓ not enough information 1

3 A trapezium has two parallel sides of lengths 9 cm and 5 cm. The perpendicular distance between those two sides is 3 cm. What is the area of the trapezium?

Ⓐ 21 cm^2 Ⓑ 30 cm^2 Ⓒ 36 cm^2 Ⓓ not enough information 1

4 What is the perimeter of the trapezium in question 3?

Ⓐ 20 cm Ⓑ 28 cm Ⓒ 34 cm Ⓓ not enough information 1

5 The longer diagonal of a kite is 1.4 m and the shorter one is 95 cm. What is the area of the kite?

Ⓐ 133 cm^2 Ⓑ 665 cm^2 Ⓒ 6650 cm^2 Ⓓ 1.33 m^2 1

6 The area of this block of land is:

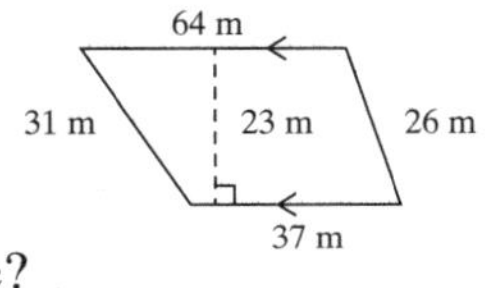

Ⓐ 1565.5 m^2 Ⓑ 1472 m^2 Ⓒ 1054.5 m^2 Ⓓ 1161.5 m^2 1

7 The diagonals of a square are both 13 cm long. What is the area of the square?

Ⓐ 42.35 cm^2 Ⓑ 84.5 cm^2 Ⓒ 169 cm^2 Ⓓ not enough information 1

8 What is the area of this shape?

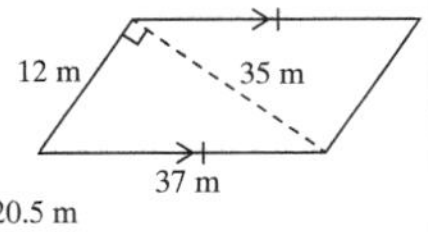

Ⓐ 420 m^2 Ⓑ 444 m^2 Ⓒ 210 m^2 Ⓓ 1295 m^2 1

9 This trapezium has perimeter 75.2 m. What is its area?

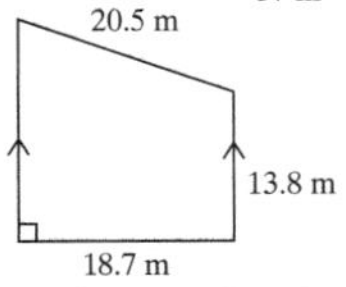

Ⓐ 270.48 m^2 Ⓑ 2.8 m^2 Ⓒ 369 m^2 Ⓓ 336.6 m^2 1

10 The length of each diagonal of a kite is known. What can be found from this information?

Ⓐ both area and perimeter Ⓑ perimeter but not area

Ⓒ area but not perimeter Ⓓ neither area nor perimeter 1

11 What is the shaded area?

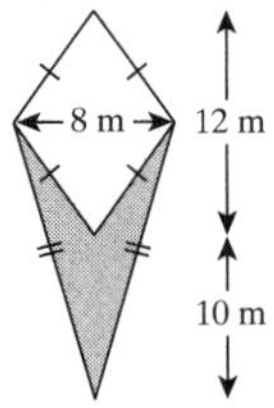

Ⓐ 40 m^2 Ⓑ 48 m^2 Ⓒ 8 m^2 Ⓓ 80 m^2 1

12 One diagonal of a rhombus is twice the length of the other. If the length of the shorter diagonal is d m, then the area of the rhombus in square metres will be:

Ⓐ $\frac{1}{2}d^2$ Ⓑ d^2 Ⓒ $2d^2$ Ⓓ $4d^2$ 1

13 The area of this shape is 120 cm^2. What is the perimeter?

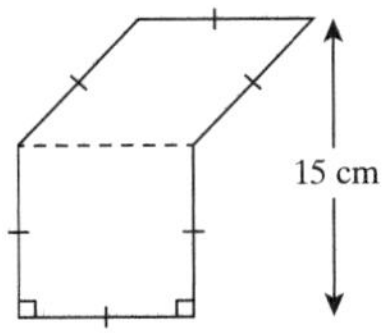

Ⓐ 48 cm Ⓑ 56 cm Ⓒ 64 cm Ⓓ 96 cm 1

14 One diagonal of a rhombus is 23 mm long and the other is 2 cm. What is the area of the rhombus?

Ⓐ 2.3 mm^2 Ⓑ 23 mm^2 Ⓒ 2.3 cm^2 Ⓓ 23 cm^2 1

15 How many hectares are in 250 km^2?

Ⓐ 2500 Ⓑ 25 000 Ⓒ 250 000 Ⓓ 2 500 000 1

PART B Write answers only in the space provided.

16 375 000 square metres is how many hectares? ______________ [1]

17 What is the length, in centimetres, of each side of a rhombus if the perimeter is 1.96 m?

______________ [1]

18 A rhombus has an area of 60 cm². If one diagonal is 12 cm long, what is the length of the other?

______________ [1]

19 How many square millimetres are in 0.5 cm²? ______________ [1]

20 The perimeter of this shape is 3.15 km.
What is the length of each of the equal sides? ______________ [1]

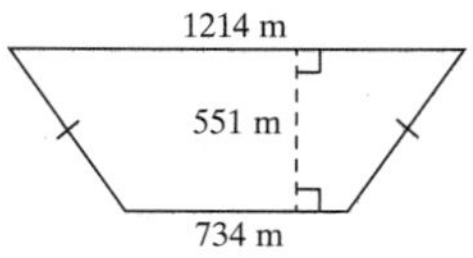

21 What is the area of the shape in question 20? Give the answer in hectares correct to one decimal place. ______________ [1]

22 What is the area of a square of perimeter 76 cm? ______________ [1]

23 A kite has an area of 72 cm². One diagonal is 16 cm long. How much shorter is the other diagonal?

______________ [1]

24 One side of a rhombus is 18 cm. The perpendicular distance between opposite sides is 13 cm.
What is the area of the rhombus? ______________ [1]

25 How long is the longest side of a rectangle, if the area is 48 cm² and the perimeter 38 cm? ______________ [1]

PART C Show all working.

26 A pattern is formed with string. It consists of five rhombuses, all with side length 901 mm, and a rectangle.

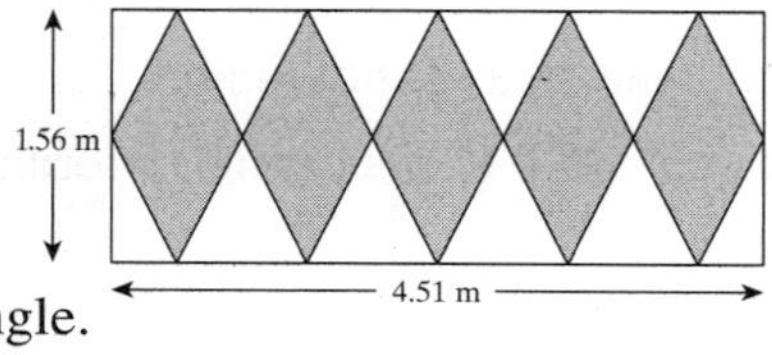

a What is the shaded area in square metres? [2 marks]

____

____

b How much string is needed? Give the answer in metres. [1 mark]

____

27 This is a regular hexagon of side length 15 cm.
What is its area? [2 marks]

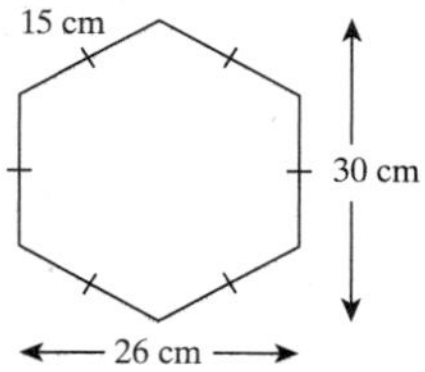

____

____

____

28 What is the area of this block of land? [2 marks]

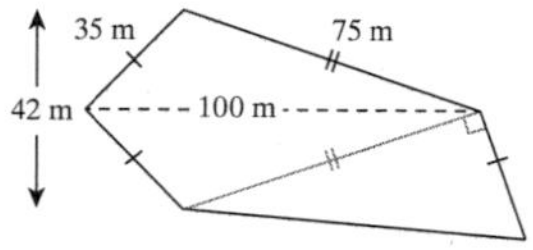

____

____

29 a A kite has diagonals of length 18 cm and 11 cm. What is its area in square metres? [1 mark]

____

b The kite has perimeter of 43.2 cm. If one of the sides is 7 cm longer than one of the others, how long is the shorter side? [2 marks]

____

____

30 This shape is made up of a right-angled triangle and a parallelogram.

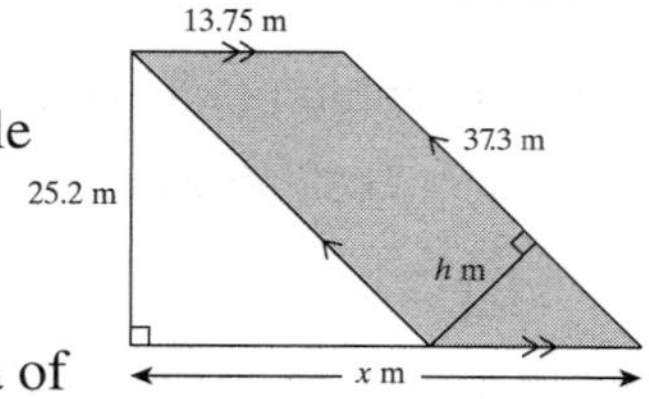

a Explain why the area of the parallelogram is 346.5 m². [1 mark]

____

b What is the value of h? (Give the answer to one decimal place.) [2 marks]

____

____

c The area of the parallelogram is half that of the whole shape. What is the value of x? [2 marks]

____

____

Total marks /40

PRETEST

Time allowed: 15 minutes **Total marks: 15**

1 A line across a circle through the middle point is called: [1]

(A) a radius (B) a diameter (C) a sector (D) an arc

2 How many degrees are in a revolution? [1]

(A) 90 (B) 180 (C) 360 (D) 380

3 The special name usually given to the point O is the: [1]

(A) middle (B) midpoint (C) centre (D) circumference

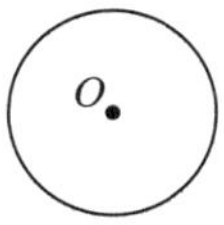

4 How many lines of symmetry does a circle have? [1]

(A) 0 (B) 1 (C) 4 (D) more than 4

5 What fraction of this circle is shaded? [1]

(A) $\frac{1}{3}$ (B) $\frac{1}{4}$ (C) $\frac{1}{5}$ (D) $\frac{1}{6}$

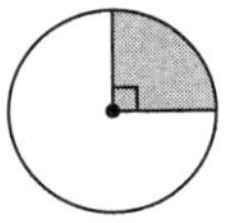

6 A line from the middle point to the outside of a circle is called: [1]

(A) an arc (B) a radius (C) a diameter (D) the circumference

7 The shaded part is called: [1]

(A) an arc (B) a sector (C) a segment (D) a chord

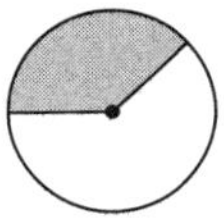

8 Half a circle is most commonly called a: [1]

(A) halfcircle (B) demicircle (C) semicircle (D) circumcircle

9 The shaded part is called: [1]

(A) an arc (B) a chord (C) a sector (D) a segment

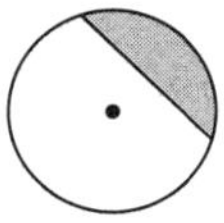

10 If the radius of a circle is 12 cm long then the diameter will be: [1]

(A) 6 cm (B) 12 cm (C) 24 cm (D) 38 cm

11 The outside of a circle is called the: [1]

(A) circumference (B) chord (C) radius (D) diameter

12 The name given to the line across the circle, as shown by the arrow, is: [1]

(A) radius (B) diameter (C) circumference (D) chord

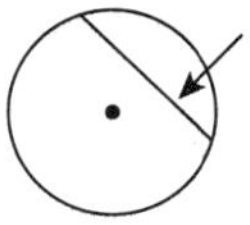

13 If the diameter is 46 cm, the radius will be: [1]

(A) 23 cm (B) 46 cm (C) 92 cm (D) 15 cm

14 The black part of this circle is: [1]

(A) an arc (B) a sector (C) a segment (D) a chord

15 What is the value of x? [1]

(A) 110 (B) 270 (C) 280 (D) 290

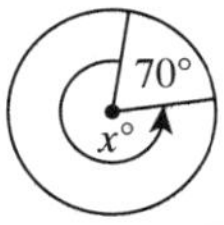

Total marks /15

Circles Calculator Allowed

LEVEL 1 TEST PART A

Time allowed: 40 minutes

1 The value of π is closest to:

(A) 2 (B) 3 (C) 4 (D) 5

1

2 If the radius of a circle is 87 cm, the diameter will be:

(A) 435 mm (B) 87 mm (C) 17.4 mm (D) 1.74 m

1

3 A circle has diameter 5 cm. Its circumference is closest to:

(A) 15 cm (B) 16 cm (C) 31 cm (D) 32 cm

1

4 A circle has radius 8 cm. Its circumference is closest to:

(A) 25 cm (B) 26 cm (C) 50 cm (D) 51 cm

1

5 The area of a circle of radius 20 cm is closest to:

(A) 126 cm^2 (B) 2513 cm^2 (C) 5027 cm^2 (D) 1257 cm^2

1

6 The circumference of this circle is closest to:

(A) 21 cm (B) 22 cm (C) 43 cm (D) 44 cm

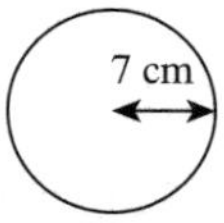

1

7 The radius of a circle is 6 cm. Its circumference will be:

(A) 3π cm (B) 6π cm (C) 12π cm (D) 24π cm

1

8 The area of this circle is closest to:

(A) 2640 mm^2 (B) 10 570 mm^2 (C) 5280 mm^2 (D) 21 140 mm^2

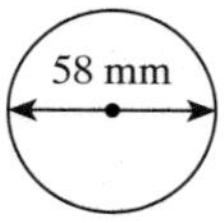

1

9 The area of a circle is 64 cm^2. What is the area of a semicircle with the same radius?

(A) 32 cm^2 (B) 16 cm^2 (C) 8 cm^2 (D) 128 cm^2

1

10 The area of this circle is closest to:

(A) 508 cm^2 (B) 254 cm^2 (C) 1018 cm^2 (D) 2036 cm^2

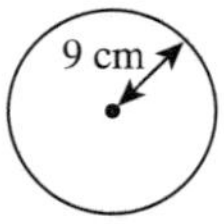

1

11 The circumference of this circle is closest to:

(A) 6038 m (B) 3019 m (C) 195 m (D) 97 m

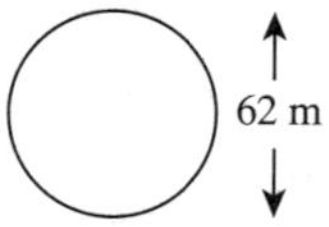

1

12 The circumference of a circle is 141 cm. Its diameter will be closest to:

(A) 45 cm (B) 22 cm (C) 24 cm (D) 12 cm

1

13 What is the area of this semicircle to the nearest square metre?

(A) 39 m^2 (B) 31 m^2 (C) 79 m^2 (D) 314 m^2

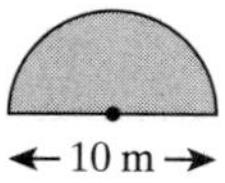

1

14 The area of this circle is 576 cm^2. What is the area of the shaded part?

(A) 144 cm^2 (B) 192 cm^2 (C) 183 cm^2 (D) 230 cm^2

1

15 What is the shaded area to the nearest square centimetre?

(A) 19 cm^2 (B) 452 cm^2 (C) 113 cm^2 (D) 226 cm^2

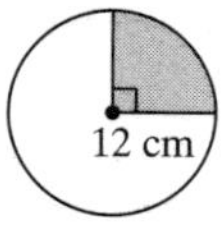

1

Circles Calculator Allowed **(Level 1 Test)**

PART B Write answers only in the space provided.

16 What is the diameter of a circle of radius 9.7 cm? ______________ 1

17 What is the value of π to five decimal places? ______________ 1

18 What is the circumference, in terms of π, of a circle of diameter 8 cm? ______________ 1

19 What is the area, in terms of π, of a circle of radius 3 cm? ______________ 1

20 What is the circumference of a circle of diameter 3 m?
(Give the answer correct to one decimal place.) ______________ 1

21 What is the area, to the nearest square centimetre, of a circle of radius 26 cm? ______________ 1

22 What is the circumference, to the nearest centimetre, of a circle of radius 47 cm? ______________ 1

23 What is the area of a circle of diameter 72 km?
(Give the answer to one decimal place.) ______________ 1

24 What is the circumference of this circle?
Give the answer to the nearest metre. ______________ 1

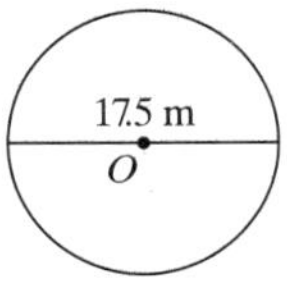

25 What is the area, to the nearest square metre,
of the circle in question 24? ______________ 1

PART C Show all working.

26 a What fraction of the full circle is shaded? 1 mark

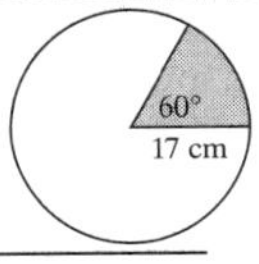

____

b What is the area, to the nearest square centimetre, of the shaded part? 1 mark

____

____

c What is the perimeter, to the nearest centimetre, of the shaded part? 2 marks

____

____

27 The shaded region is the area between two circles.
What is the area in square centimetres to one decimal place? 2 marks

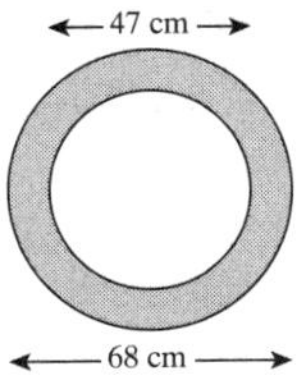

____

____

28 a A circular running track has a radius of 64 m. How many whole laps of the track would you need to complete to run 2 km? 2 marks

____

____

b How much further is it, to the nearest metre, around the track from one side to the other than straight across the centre? 2 marks

____

____

29 This shape is made up of a rectangle and semicircle.

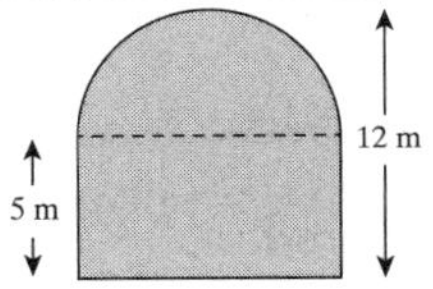

a Briefly explain why the length of the rectangle is 14 m. 1 mark

____

____

____

b What is the area of the shape to the nearest square metre? 2 marks

____

____

c What is the perimeter of the shape to the nearest metre? 2 marks

____

____

Total marks /40

Circles Calculator Allowed

LEVEL 2 TEST — PART A

Time allowed: 40 minutes

1 A circle has diameter 83 cm. What is its circumference, to the nearest metre?
(A) 1 m (B) 2 m (C) 3 m (D) 4 m — 1

2 A circle has radius 19 m. What is its area to the nearest square metre?
(A) 119 m^2 (B) 1134 m^2 (C) 4536 m^2 (D) 2268 m^2 — 1

3 A circle has a diameter of 16 cm. What is its area in terms of π?
(A) 64π cm^2 (B) $64\pi^2$ cm^2 (C) 8π cm^2 (D) $8\pi^2$ cm^2 — 1

4 The circumference of a circle of radius 23 cm is closest to:
(A) 72 cm (B) 415 cm (C) 145 cm (D) 289 cm — 1

5 The area of a semicircle of radius 9.5 m is closest to:
(A) 142 m^2 (B) 60 m^2 (C) 284 m^2 (D) 567 m^2 — 1

6 The circumference of a circle is 9π cm. What is the radius?
(A) 3 cm (B) 4.5 cm (C) 9 cm (D) 18 cm — 1

7 A circle has diameter 2.3 m. What is its area in square metres to two decimal places?
(A) 4.15 m^2 (B) 7.23 m^2 (C) 14.45 m^2 (D) 16.62 m^2 — 1

8 The circumference of a circle of radius 750 m is closest to:
(A) 2.4 km (B) 4.7 km (C) 1.8 km (D) 3.5 km — 1

9 About how much longer is the circumference than the diameter of a circle of radius 11 cm?
(A) 24 cm (B) 13 cm (C) 47 cm (D) 58 cm — 1

10 The area of a circle is 36 cm^2. Its radius is closest to:
(A) 2 cm (B) 3 cm (C) 4 cm (D) 5 cm — 1

11 The area, to the nearest square centimetre, of a quadrant of a circle of radius 17 cm is:
(A) 227 cm^2 (B) 107 cm^2 (C) 454 cm^2 (D) 27 cm^2 — 1

12 As shown in the diagram, semicircles are drawn on each side of a square of side 6 cm. What is the shaded area in square centimetres to one decimal place?

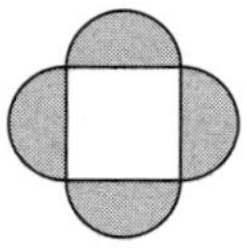

(A) 28.3 cm^2 (B) 75.4 cm^2 (C) 226.2 cm^2 (D) 56.5 cm^2 — 1

13 How many times greater is the circumference of a circle of diameter 32 m than one of radius 2 m?
(A) 4 (B) 8 (C) 16 (D) 30 — 1

14 This circle has a radius of 28 cm. The shaded area is closest to:

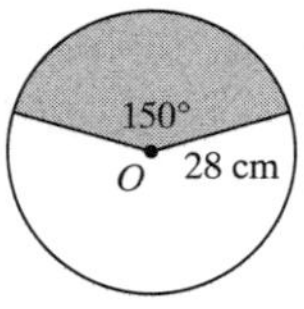

(A) 411 cm^2 (B) 274 cm^2 (C) 493 cm^2 (D) 1026 cm^2 — 1

15 The diagram shows a quadrant of a circle of radius 32 cm. Its perimeter is closest to:

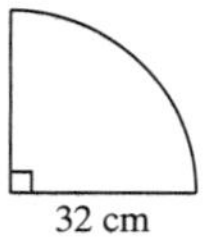

(A) 50.3 cm (B) 132.5 cm (C) 82.3 cm (D) 114.3 cm — 1

Circles Calculator Allowed **(Level 2 Test)**

PART B Write answers only in the space provided.

16 The diameter of a circle is 37.8 cm. What is the radius? ________________ [1]

17 What is the circumference, in terms of π, of a circle of radius 1.5 cm? ________________ [1]

18 What is the area of a circle, in terms of π, if the radius is 12 cm? ________________ [1]

19 What is the value of π to seven decimal places? ________________ [1]

20 What is the circumference of a circle of radius 2.5 m?
Give the answer to two decimal places. ________________ [1]

21 What is the area of a circle with diameter 49 cm? Give the answer to one decimal place. ____________ [1]

22 A circle has circumference 20π cm. What is the diameter? ________________ [1]

23 What is the area of this semicircle?
Give the answer to one decimal place. ________________ [1]

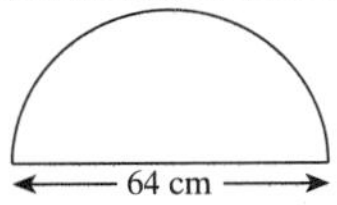

24 A circle has circumference 12.6 m. What is the diameter to the nearest metre? ________________ [1]

25 The area of a circle is 4778 cm^2. What is the radius to the nearest centimetre? ________________ [1]

PART C Show all working.

26 A circle has circumference 21 cm.

a What is its diameter to the nearest millimetre? [2 marks]

____

____

b What is its area to the nearest square centimetre? [2 marks]

____

____

27 Two circles fit into a rectangle as shown.

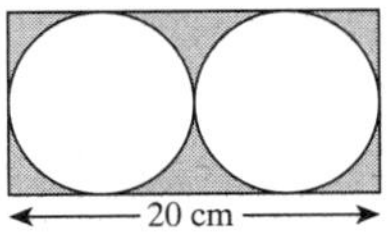

a What is the width of the rectangle? [1 mark]

____

b What is the shaded area to one decimal place? [2 marks]

____

____

____

____

28 A sporting field is made up of a rectangle and two semicircles as shown in the diagram.

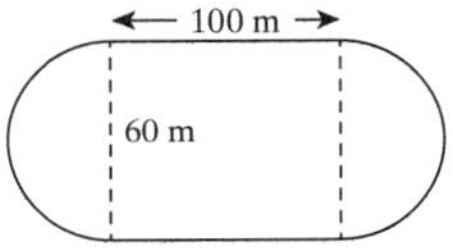

a What is the area of the field to the nearest square metre? [2 marks]

____

____

b What is the perimeter of the field to the nearest metre? [2 marks]

____

____

29 a What is the area of the shaded region? (Give the answer to one decimal place.) [2 marks]

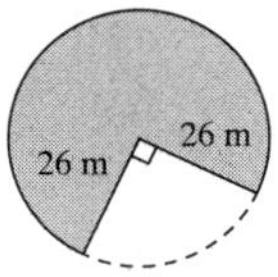

____

____

b What is the perimeter of the shaded region? (Give the answer to one decimal place.) [2 marks]

____

____

Total marks /40

Circles Calculator Allowed

LEVEL 3 TEST PART A

Time allowed: 40 minutes

1 Which is closest to the value of π?

(A) $\frac{22}{7}$ (B) $\frac{16}{5}$ (C) $\frac{19}{6}$ (D) $\frac{25}{8}$ 1

2 A circle has radius 11 cm. What is its circumference to the nearest millimetre?

(A) 346 mm (B) 345 mm (C) 691 mm (D) 692 mm 1

3 The area of a circle of radius 95 cm is closest to:

(A) 2.8 m^2 (B) 3.0 m^2 (C) 6.0 m^2 (D) 5.7 m^2 1

4 What is the radius of a circle of area 25π cm^2?

(A) 12.5 cm (B) 10 cm (C) 5 cm (D) 25 cm 1

5 The area of a semicircle of radius 15 cm is closest to:

(A) 707 cm^2 (B) 177 cm^2 (C) 188 cm^2 (D) 353 cm^2 1

6 Which is closest to the shaded area?

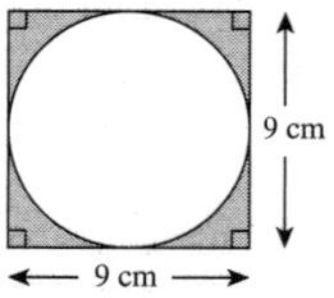

(A) 17.4 cm^2 (B) 52.7 cm^2 (C) 173.5 cm^2 (D) 63.6 cm^2 1

7 The circumference of a circle is 48 cm. Its diameter is closest to:

(A) 8 cm (B) 15 cm (C) 14 cm (D) 31 cm 1

8 The perimeter of this semicircle is closest to:

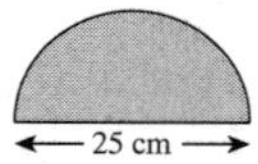

(A) 39 cm (B) 64 cm (C) 78 cm (D) 1.04 m 1

9 The area of a circle is 856 cm^2. What is its diameter closest to?

(A) 17 cm (B) 18 cm (C) 23 cm (D) 33 cm 1

10 The circumference of a circle is 18 cm. What is its approximate area?

(A) 103.1 cm^2 (B) 81 cm^2 (C) 32.8 cm^2 (D) 25.8 cm^2 1

11 The shaded area is approximately 531 cm^2. What is the radius of the circle?

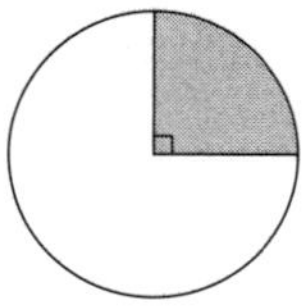

(A) 26 cm (B) 13 cm (C) 52 cm (D) 15 cm 1

12 The area of a circle is 623 cm^2. Its circumference is closest to:

(A) 50 cm (B) 45 cm (C) 25 cm (D) 90 cm 1

13 The diagram shows a rectangle and semicircle. What is the area to the nearest square centimetre?

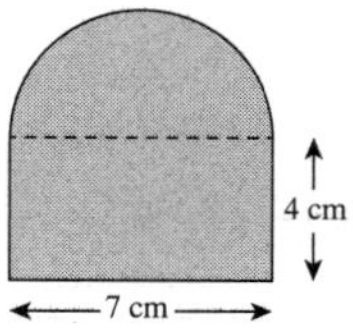

(A) 39 cm^2 (B) 47 cm^2 (C) 66 cm^2 (D) 105 cm^2 1

14 What is the perimeter, to the nearest centimetre, of the figure in question 13?

(A) 37 cm (B) 22 cm (C) 26 cm (D) 33 cm 1

15 The circumference (in metres) and the area (in square metres) of a circle have the same magnitude. What is the radius of the circle?

(A) 1 m (B) 2 m (C) π m (D) 4 m 1

Circles Calculator Allowed **(Level 3 Test)**

PART B Write answers only in the space provided.

16 What is the value of 2π to five decimal places? ________________ [1]

17 What is the circumference of a circle of radius 14 cm? Give the answer in terms of π. ____________ [1]

18 Find the area, in terms of π, of a circle of diameter 18 cm. ________________ [1]

19 What is the circumference, in metres, of a circle of diameter 0.87 m? Give the answer to the nearest centimetre. ________________ [1]

20 What is the area of a circle of radius 93 cm? Give the answer in square metres correct to two decimal places. ________________ [1]

21 What is the diameter, in centimetres to the nearest centimetre, of a circle of circumference 2.1 m? ________________ [1]

22 What is the area of a quadrant of a circle of radius 16 cm? ________________ [1]

23 What is the radius of a circle of area 1662 cm^2? ________________ [1]

24 What is the area of a circle, in terms of π, if the circumference is 38π cm? ________________ [1]

25 What is the circumference, to the nearest centimetre, of a circle of area 113 cm^2? ________________ [1]

PART C Show all working.

26 This shape is made up of two semicircles.

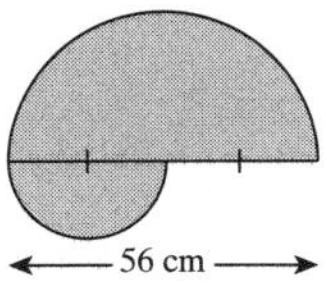

a What is the area of the shape? [2 marks]

____

b What is the perimeter of the shape? [2 marks]

____

27 A square fits exactly into a circle of diameter 30 cm.

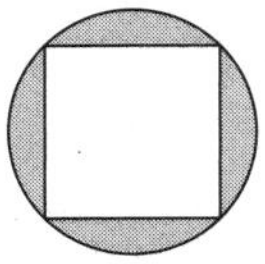

What is the shaded area? [3 marks]

____

28 The shaded area is part of the area between two circles.

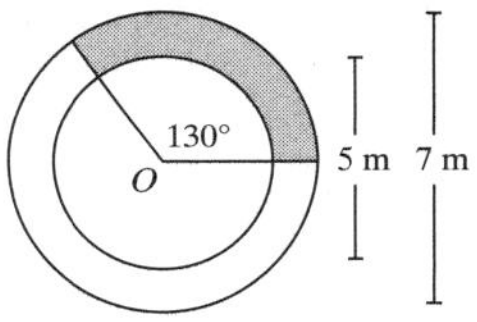

a What is the area of the shaded part? [2 marks]

____

b What is the perimeter of the shaded part? [2 marks]

____

29 This shaded region is bound by the arcs of two quadrants.

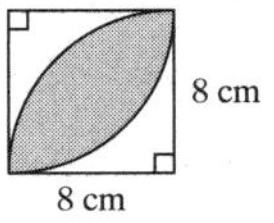

a Find the perimeter of the shaded region. (Give the answer to one decimal place.) [2 marks]

____

b Find the shaded area. (Give the answer to one decimal place.) [2 marks]

____

Total marks /40

PRETEST

Time allowed: 15 minutes **Total marks: 15**

1 Which would be the smallest volume?

(A) 0.6 km^3 (B) 5 m^3 (C) 27 cm^3 (D) 320 mm^3 [1]

2 Which unit would be the most appropriate to measure the volume of a shipping container?

(A) mm^3 (B) cm^3 (C) m^3 (D) km^3 [1]

3 Which prism has the greatest volume?

(A) 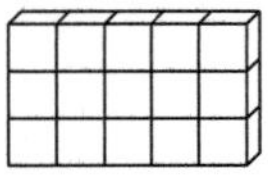 (B) 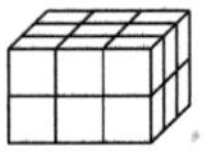 (C) 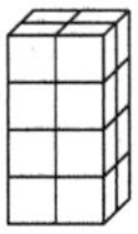 (D) 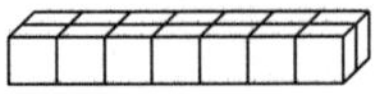 [1]

4 How many millilitres are in 3.07 L?

(A) 3007 (B) 3070 (C) 370 (D) 307 [1]

5 Which jug holds the most juice?

(A) 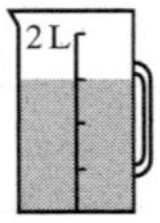

(B) 

(C) 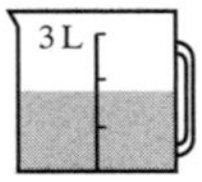

(D) 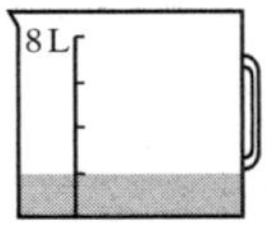

[1]

6 The amount of water in a normal plastic bucket, when full, would be closest to:

(A) 700 mL (B) 70 L (C) 7 L (D) 7 kL [1]

7 What is the area of this triangle?

(A) 1470 cm^2 (B) 2626 cm^2 (C) 2940 cm^2 (D) 3822 cm^2 [1]

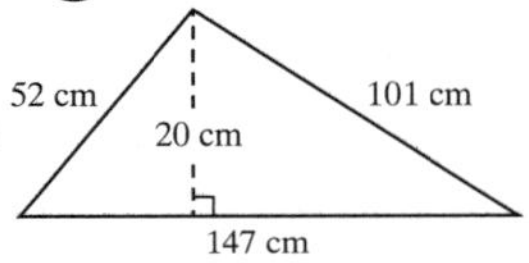

8 What would be the volume of a cube of side 4 cm?

(A) 4 cm^3 (B) 16 cm^3 (C) 24 cm^3 (D) 64 cm^3 [1]

9 What is the area of a rectangle 1.2 m long and 90 cm wide?

(A) 108 m^2 (B) 10.8 m^2 (C) 1.08 m^2 (D) 1080 cm^2 [1]

10 What would be the volume of this prism in cubic units?

(A) 11 (B) 24 (C) 26 (D) 36 [1]

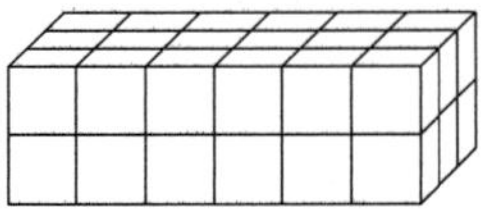

11 If $V = Ah$, what is the value of V when $A = 7$ and $h = 3$?

(A) 10 (B) 21 (C) 73 (D) 343 [1]

12 What is the area, in square metres to two decimal places, of a square of side length 875 mm?

(A) 0.77 m^2 (B) 7.65 m^2 (C) 79.56 m^2 (D) 765.65 m^2 [1]

13 How many bricks would there be in a stack 3 wide, 4 long and 2 high?

(A) 9 (B) 12 (C) 20 (D) 24 [1]

14 A rectangle with which of these dimensions would have a perimeter of 28 cm and area 48 cm^2?

(A) 12 cm by 4 cm (B) 16 cm by 3 cm (C) 8 cm by 6 cm (D) 7 cm by 7 cm [1]

15 What is the area of a right-angled triangle with sides 14 cm, 17.1 cm and 22.1 cm?

(A) 119.7 cm^2 (B) 154.7 cm^2 (C) 239.4 cm^2 (D) 188.955 cm^2 [1]

Total marks /15

Volume Calculator Allowed

LEVEL 1 TEST — PART A

Time allowed: 40 minutes

1 A rectangular prism has base area 24 m^2 and perpendicular height 5 m. What is its volume?

(A) 29 m^2 (B) 60 m^3 (C) 120 m^3 (D) 129 m^3 [1]

2 How many millilitres are in 7.3 L?

(A) 730 (B) 7030 (C) 7300 (D) 73 000 [1]

3 A container with volume 20 cm^3 will have a capacity of:

(A) 2 mL (B) 20 mL (C) 200 mL (D) 20 L [1]

4 What is the volume of this triangular prism?

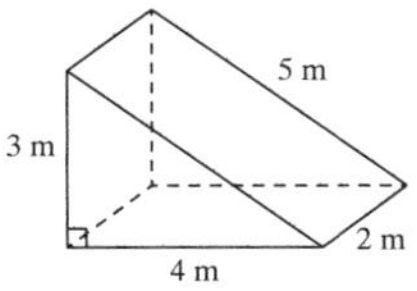

(A) 12 m^3 (B) 24 m^3 (C) 30 m^3 (D) 60 m^3 [1]

5 How many cubic millimetres are in 1 cm^3?

(A) 10 (B) 1000 (C) 10 000 (D) 100 000 [1]

6 What is the capacity of this rectangular prism in millilitres?

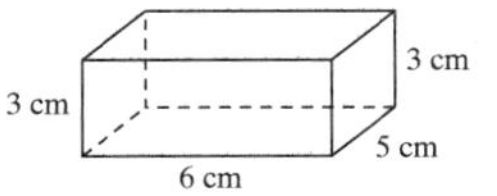

(A) 9 mL (B) 27 mL (C) 54 mL (D) 90 mL [1]

7 A rectangular prism with which dimensions will have the greatest volume?

(A) 13 cm, 2 cm, 1 cm (B) 7 cm, 2 cm, 2 cm

(C) 5 cm, 3 cm, 2 cm (D) 3 cm, 3 cm and 3 cm [1]

8 How many cubic centimetres are in 1 m^3?

(A) 100 (B) 1000 (C) 10 000 (D) 1 000 000 [1]

9 What is the volume of a cube of side 9 m?

(A) 81 m^3 (B) 486 m^3 (C) 729 m^3 (D) 19 683 m^3 [1]

10 The volume of which prism is found by multiplying the area of the shaded face by *h*?

(A) 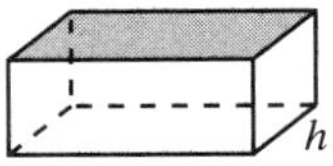

(B) 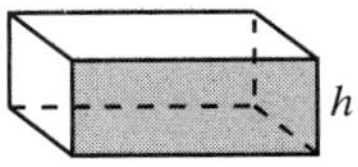

(C) 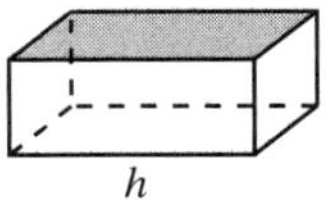

(D) 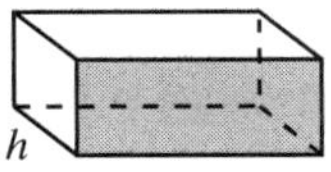

[1]

11 A container has a capacity of 3 kL. What is its volume?

(A) 3 m^3 (B) 30 m^3 (C) 300 m^3 (D) 3000 m^3 [1]

12 A rectangular prism is 3.2 m long, 2.4 m wide and 0.9 m high. Its volume is closest to:

(A) 6 m^3 (B) 7 m^3 (C) 8 m^3 (D) 9 m^3 [1]

13 What is the volume of this prism?

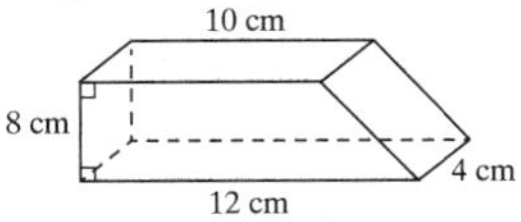

(A) 352 cm^3 (B) 320 cm^3 (C) 384 cm^3 (D) 720 cm^3 [1]

14 What is the volume, to the nearest cubic centimetre, of a cylinder of radius 4 cm and height 6 cm?

(A) 302 cm^3 (B) 1206 cm^3 (C) 452 cm^3 (D) 151 cm^3 [1]

15 What is the volume of concrete in this block?

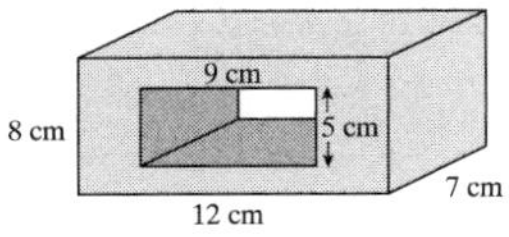

(A) 312 cm^3 (B) 357 cm^3 (C) 476 cm^3 (D) 627 cm^3 [1]

Volume Calculator Allowed **(Level 1 Test)**

PART B Write answers only in the space provided.

16 What is the volume of a cube of side 10 cm? ________ 1

17 How many litres are in $2\frac{1}{2}$ kL? ________ 1

18 A container with volume 800 cm^3 will have what capacity? ________ 1

19 What is the volume of a rectangular prism 9 cm long, 6 cm wide and 4 cm high? ________ 1

20 How many litres is 15 600 mL? ________ 1

21 How many cubic centimetres is 60 m^3? ________ 1

22 The area of the cross-section of a prism is 2.7 m^2 and the perpendicular height is 0.8 m. What is the volume of the prism? ________ 1

23 What is the volume of a cylinder of radius 6 cm and height 8 cm? (Give the answer in cubic centimetres correct to one decimal place.) ________ 1

24 How many litres are in a megalitre? ________ 1

25 The area of the shaded face is 30 cm^2. What is the volume? ________ 1

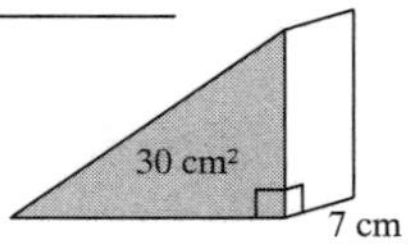

PART C Show all working.

26 A fish tank, a rectangular prism, holds 24 L of water.

a How many millilitres does it hold? 1 mark

b What would be the volume in cubic centimetres? 1 mark

c If the tank is 40 cm long and 30 cm wide, how deep must the water be? 2 marks

27 A skip bin is in the shape of a trapezoidal prism.

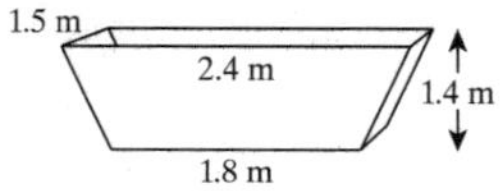

What is the volume of the bin in cubic metres? 2 marks

28 a What is the volume of this triangular prism in cubic millimetres? 2 marks

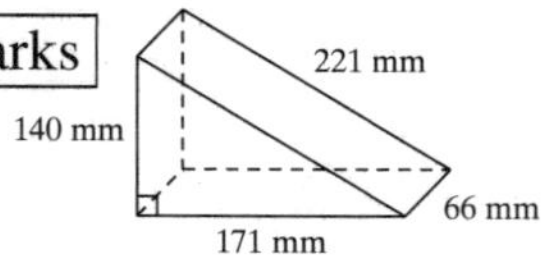

b What is the volume in cubic centimetres, to the nearest cubic centimetre? 1 mark

29 The diagram shows a driveway that is to be concreted.

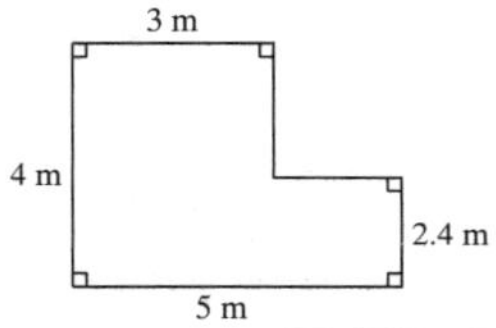

a What is the area of the driveway? 1 mark

b If the concrete will be laid to a depth of 10 cm, how many cubic metres of concrete will be needed? 2 marks

30 a What is the volume of this cylinder to the nearest cubic metre? 2 marks

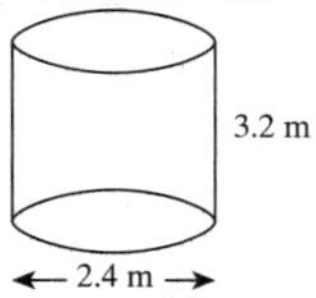

b How many litres of water will it hold, to the nearest thousand litres, when full? 1 mark

Total marks /40

LEVEL 2 TEST PART A

Time allowed: 40 minutes

1 A rectangular prism is 1.2 m long, 80 cm wide and 60 cm high. What is its volume?

Ⓐ 0.005 76 m^3 Ⓑ 0.576 m^3 Ⓒ 5.76 m^3 Ⓓ 5760 m^3 1

2 A container has a capacity of 600 mL. What is its volume?

Ⓐ 600 mm^3 Ⓑ 6 cm^3 Ⓒ 60 cm^3 Ⓓ 600 cm^3 1

3 How many litres are in 3.06 kL?

Ⓐ 3006 Ⓑ 306 Ⓒ 30 600 Ⓓ 3060 1

4 The volume of which prism is **not** found by multiplying the area of the shaded face by x?

Ⓐ 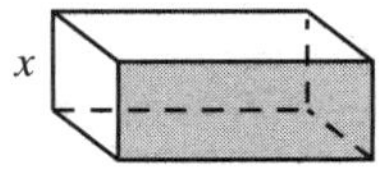

Ⓑ 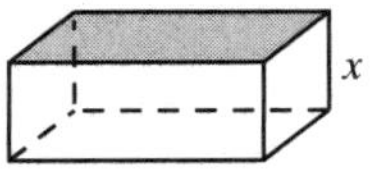

Ⓒ 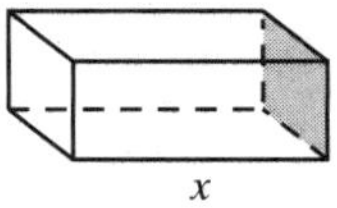

Ⓓ 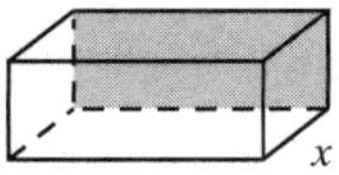

1

5 Nine cubic metres is how many cubic millimetres?

Ⓐ 9 000 000 000 Ⓑ 9 000 000 Ⓒ 90 000 Ⓓ 9000 1

6 A cube has volume 4096 cm^3. What is the length of each side?

Ⓐ 64 cm Ⓑ 36 cm Ⓒ 24 cm Ⓓ 16 cm 1

7 What is the volume of this triangular prism?

Ⓐ 312 cm^3 Ⓑ 480 cm^3 Ⓒ 520 cm^3 Ⓓ 780 cm^3 1

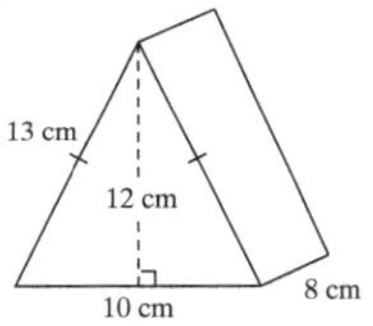

8 These are the dimensions (in centimetres) of some rectangular prisms. Which does **not** have volume 25 200 cm^3?

Ⓐ 60 × 28 × 15 Ⓑ 63 × 25 × 16 Ⓒ 70 × 40 × 9 Ⓓ 72 × 35 × 8 1

9 The constant cross-section of this prism is a parallelogram. What is the volume of this prism?

Ⓐ 2142 cm^3 Ⓑ 3780 cm^3 Ⓒ 4590 cm^3 Ⓓ 7140 cm^3 1

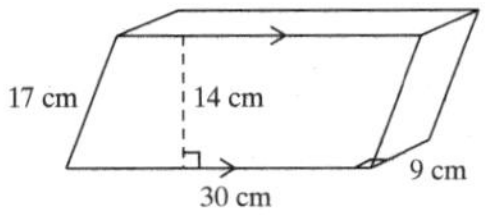

10 How many cubic centimetres are in 0.5 m^3?

Ⓐ fifty Ⓑ five thousand Ⓒ five hundred thousand Ⓓ five million 1

11 What is the capacity of this prism in litres?

Ⓐ 1.6 L Ⓑ 16 L Ⓒ 160 L Ⓓ 1600 L 1

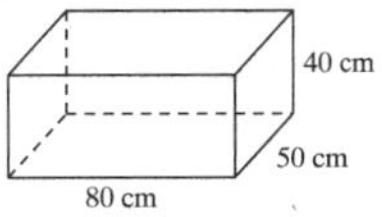

12 A triangular prism has base area 68 cm^2 and perpendicular height 12 cm. What is its volume?

Ⓐ 172 cm^3 Ⓑ 408 cm^3 Ⓒ 816 cm^3 Ⓓ 1632 cm^3 1

13 A container with a volume 1.8 m^3 will have capacity:

Ⓐ 180 L Ⓑ 1800 L Ⓒ 1.8 L Ⓓ 18 L 1

14 Which rectangular prism has the greatest volume?

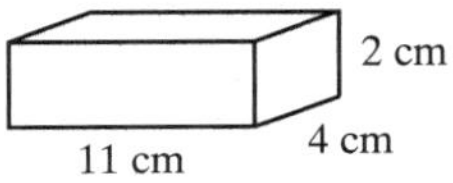

Ⓐ

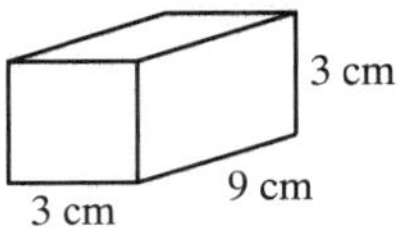

Ⓑ

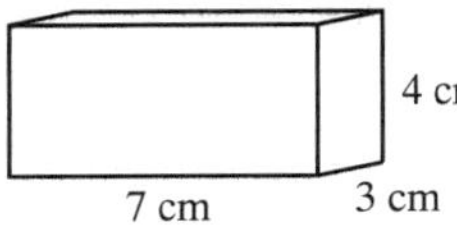

Ⓒ

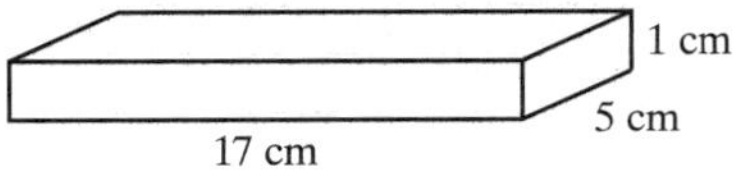

Ⓓ

1

15 What is the volume of this prism?

Ⓐ 270 cm^3 Ⓑ 310 cm^3 Ⓒ 330 cm^3 Ⓓ 360 cm^3 1

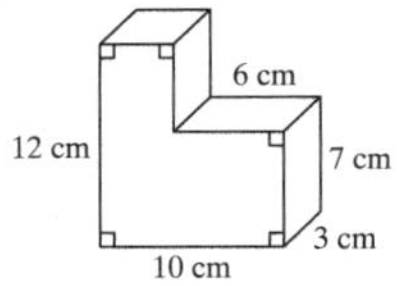

PART B Write answers only in the space provided.

16 How many millilitres are in $10\frac{1}{4}$ L? ______________ 1

17 A container with capacity of 20 L has what volume in cubic metres? ______________ 1

18 What is the volume of the prism shown on the right? ______________ 1

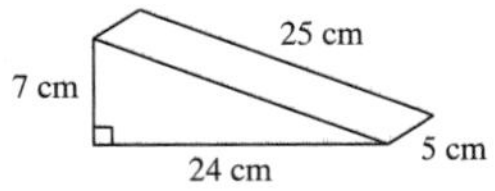

19 What is the capacity, in millilitres, of a rectangular prism 8 cm long, 6 cm wide and 5 cm high? ______________ 1

20 What is the area of the cross-section of a cube with volume 64 cm^3? ______________ 1

21 Three hundred kilolitres is how many megalitres? ______________ 1

22 A prism has perpendicular height 11 cm and the area of its cross-section is 37 cm^2. What is its volume? ______________ 1

23 What is the volume of a cylinder of radius 24 cm and height 17 cm? (Give the answer to the nearest cubic centimetre.) ______________ 1

24 What is the volume of the prism shown on the right? ______________ 1

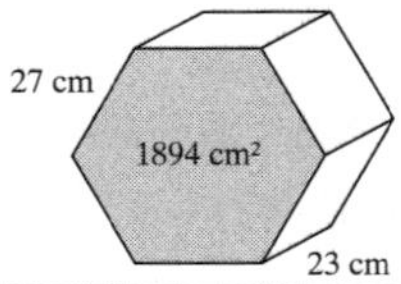

25 How many cubic millimetres are in 0.9 cm^3? ______________ 1

PART C Show all working.

26 A swimming pool has a trapezoidal cross-section as shown.

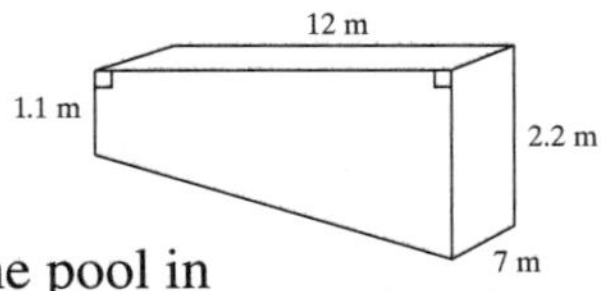

a What is the volume of the pool in cubic metres? 2 marks

b How many litres of water will the pool hold when full? 1 mark

27 Find the volume, in cubic metres, of this triangular prism. 2 marks

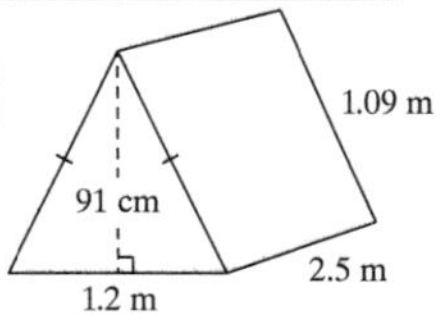

28 A cylindrical drum has a radius of 20 cm and height 70 cm.

a Find the volume of the drum to the nearest cubic centimetre. 2 marks

b How many litres will the drum hold when full? Give the answer to the nearest litre. 2 marks

29 The diagram shows a concrete block. How many of these blocks could be made from one cubic metre of concrete? 3 marks

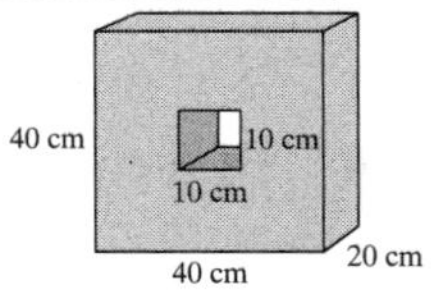

30 This diagram shows a plan for a podium to be made from a resin and used at a sporting event.

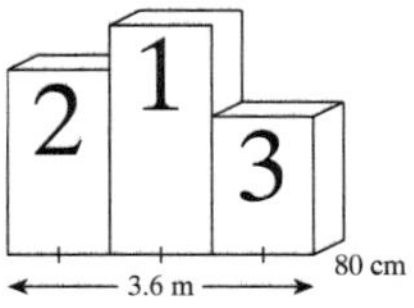

a First place will have a height of 1 m and third place will be 60 cm high. If second place is halfway between these heights, how high will it be? 1 mark

b Find the volume of resin needed to make the podium. 2 marks

Total marks /40

LEVEL 3 TEST PART A

Time allowed: 40 minutes

1 A rectangular prism is 2.1 m long, 75 cm wide and 625 mm high. Its volume would be closest to:
(A) $10\ m^3$ (B) $1\ m^3$ (C) $0.1\ m^3$ (D) $0.01\ m^3$ [1]

2 The volume of which of these prisms is found by multiplying the area of the shaded face by x?

(A) 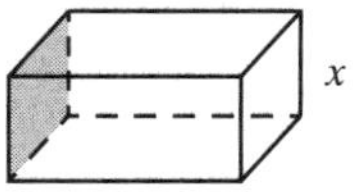

(B) 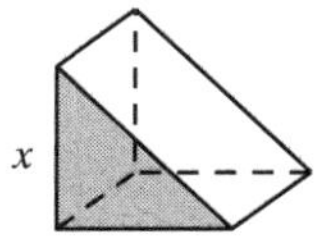

(C) 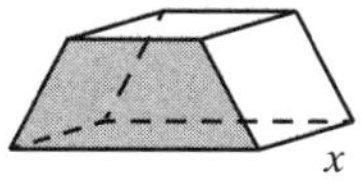

(D) 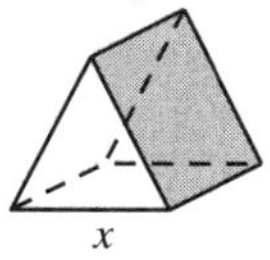

[1]

3 How many litres are in 0.04 ML?
(A) 40 000 (B) 4000 (C) 40 (D) 400 000 [1]

4 Which of these rectangular prisms will have the least volume?

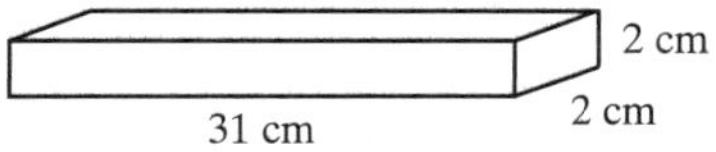

(A)
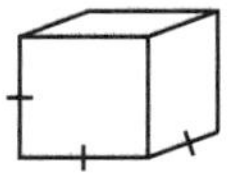

(B)
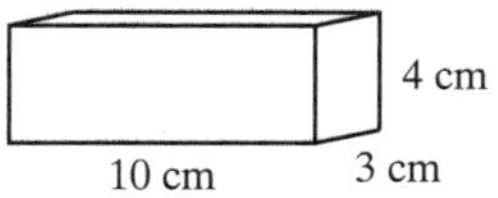

(C)
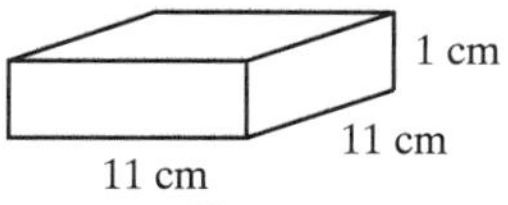

(D)
[1]

5 A container has a capacity of 40 L. What is its volume in cubic centimetres?
(A) 40 000 000 (B) 4 000 000 (C) 400 000 (D) 40 000 [1]

6 What is the volume of this triangular prism?
(A) $12\,470\ cm^3$ (B) $19\,720\ cm^3$ (C) $22\,446\ cm^3$ (D) $42\,398\ cm^3$ [1]

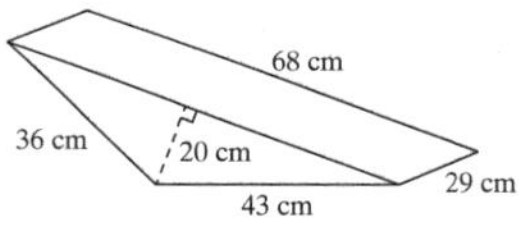

7 What is the volume of a cube of side 25 mm?
(A) $625\ mm^3$ (B) $15.625\ cm^3$ (C) $6.25\ cm^3$ (D) $1562.5\ mm^3$ [1]

8 How many cubic metres is $700\ cm^3$?
(A) 7 (B) 0.7 (C) 0.007 (D) 0.0007 [1]

9 What is the volume of this prism?
(A) $1008\ m^3$ (B) $510\ m^3$ (C) $1020\ m^3$ (D) $2016\ m^3$ [1]

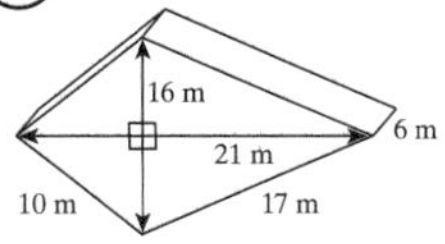

10 A rectangular prism with which dimensions, in metres, will have the greatest volume?
(A) $1.3 \times 1.3 \times 0.9$ (B) $1.8 \times 1.2 \times 0.7$ (C) $1.9 \times 1.6 \times 0.5$ (D) $2.3 \times 1.1 \times 0.6$ [1]

11 A trapezoidal prism has volume $1008\ m^3$. If the area of the cross-section is $42\ m^3$, what is the perpendicular height of the prism?
(A) 36 m (B) 24 m (C) 18 m (D) 12 m [1]

12 What is the capacity, in litres, of this prism?
(A) 0.252 (B) 2.52 (C) 25.2 (D) 252 [1]

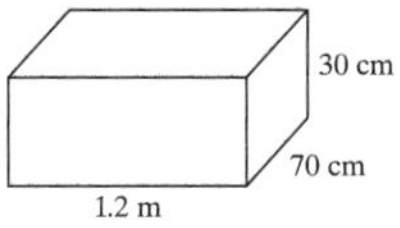

13 A container with volume $180\ m^3$ will have a capacity of:
(A) 18 ML (B) 1.8 ML (C) 0.18 ML (D) 0.018 ML [1]

14 What is the volume of this prism in cubic centimetres?
(A) 3936 (B) 3320 (C) 3276 (D) 3474 [1]

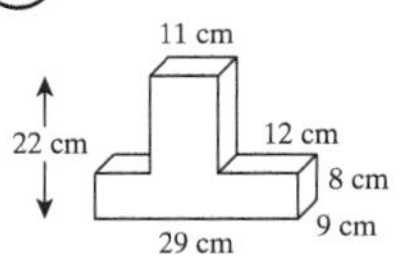

15 Which is closest to the capacity of a cylinder of diameter 30 cm and height 20 cm?
(A) 2 L (B) 4 L (C) 14 L (D) 57 L [1]

Volume Calculator Allowed **(Level 3 Test)**

PART B Write answers only in the space provided.

16 What is the length of each side of a cube of capacity 8 L? _______________ 1

17 How many kilolitres are in 0.05 ML? _______________ 1

18 What is the capacity, in litres, of a container if the area of the cross-section is 2.5 m^2 and the perpendicular height is 80 cm? _______________ 1

19 How many cubic centimetres is 20 mm^3? _______________ 1

20 What is the volume, to the nearest cubic metre, of a cylinder of diameter 3.2 m and height 1.9 m? _______________ 1

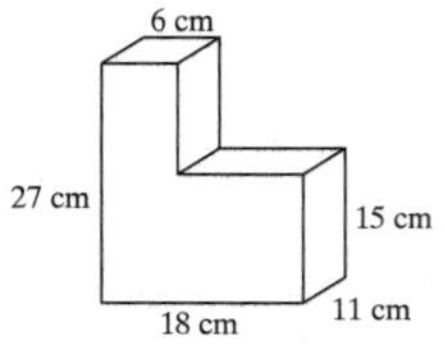

21 What is the volume of the prism shown on the right? _______________ 1

22 What is the difference in volume between a rectangular prism 64 cm by 63 cm by 3 cm and one that is 36 cm by 28 cm by 12 cm? _______________ 1

23 A cylinder of height 4 cm has volume 36π cm^3. What is the radius? _______________ 1

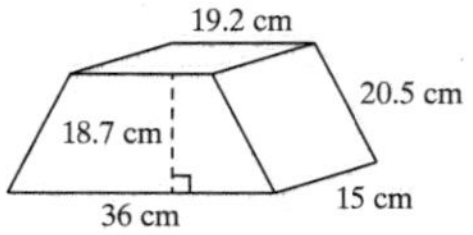

24 What is the volume of the prism shown on the right? _______________ 1

25 A right-angled triangle has sides of 39 m, 80 m and 89 m. If it forms the cross-section of a prism of height 26 m, what is the volume of the prism? _______________ 1

PART C Show all working.

26 All measurements for this triangular prism are in millimetres.

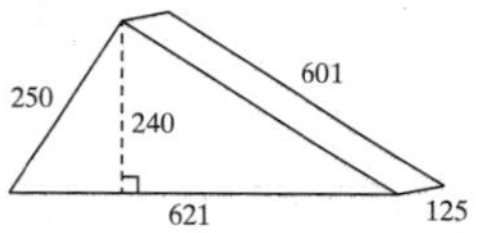

a What is the volume in cubic centimetres? [2 marks]

____

____

b What is that volume in cubic metres? [1 mark]

____

27 The diagram shows the side wall (cross-section) of a swimming pool of width 4 m.

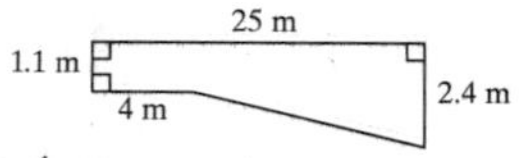

a What volume of water, in litres, will the pool hold when completely full? [3 marks]

____

____

____

b How much less water will it hold if filled to a point 15 cm from the top? [1 mark]

____

28 What is the volume of this container, in cubic centimetres? [2 marks]

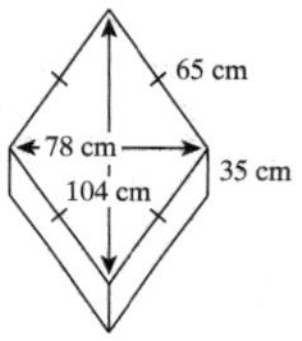

____

____

29 What is the total volume of water, in litres, in this pipe when full? [3 marks]

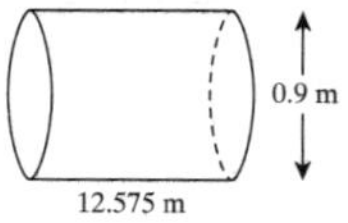

____

____

____

30 The diagram shows part of a garden made up of a square and quadrant.

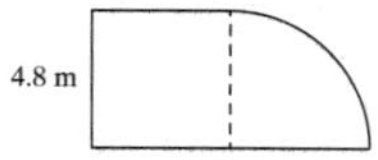

This part of the garden will be covered in mulch to an average depth of 6 cm. What will be the cost of the mulch, to the nearest dollar, at $56 per cubic metre? [3 marks]

____

____

____

Total marks /40

PRETEST

Time allowed: 15 minutes **Total marks 15**

1 How many seconds are in 5 minutes?
(A) 200 (B) 300 (C) 400 (D) 500 [1]

2 What is the time $1\frac{1}{2}$ hours after 6:45 pm?
(A) 7:15 pm (B) 7:35 pm (C) 8:15 pm (D) 8:35 pm [1]

3 Which of these months has 31 days?
(A) February (B) April (C) June (D) August [1]

4 What time is shown on this clock?
(A) 7:47 (B) 8:47 (C) 9:38 (D) 10:38

[1]

5 How many hours are in 5 days?
(A) 60 (B) 80 (C) 100 (D) 120 [1]

6 How many minutes are in $2\frac{1}{2}$ hours?
(A) 110 (B) 150 (C) 210 (D) 250 [1]

7 Which time is the same as a quarter to five?
(A) 4:15 (B) 4:45 (C) 5:15 (D) 5:45 [1]

8 How many days are in 7 weeks?
(A) 35 (B) 42 (C) 49 (D) 70 [1]

9 How long is it from 8:20 am until 3:40 pm on the same day?
(A) 5 h 20 min (B) 6 h 20 min (C) 6 h 40 min (D) 7 h 20 min [1]

10 Which clock shows a quarter past two?
(A) 2:15 (B) 2:25 (C) 2:35 (D) 2:45 [1]

11 If 10 May is a Thursday, what will 29 May be?
(A) Monday (B) Tuesday (C) Wednesday (D) Saturday [1]

12 How many seconds are in an hour?
(A) 60 (B) 360 (C) 600 (D) 3600 [1]

13 How many hours is it from 5 pm Tuesday until 7 am Thursday in the same week?
(A) 38 (B) 34 (C) 26 (D) 50 [1]

14 What is the time $\frac{3}{4}$ hour after 3:25 pm?
(A) 3:50 pm (B) 4 pm (C) 4:10 pm (D) 4:15 pm [1]

15 What is 7:25 pm in 24-hour time?
(A) 1725 (B) 1825 (C) 1925 (D) 2025 [1]

Total marks /15

LEVEL 1 TEST PART A

Time allowed: 40 minutes

1 How long is 150 minutes in hours and minutes?
(A) 1 h 50 min (B) 2 h 5 min (C) 2 h 50 min (D) 2 h 30 min [1]

2 Which is closest to the number of weeks in a year?
(A) 48 (B) 52 (C) 56 (D) 60 [1]

3 What is the date 12 days after 25 December?
(A) 4 January (B) 5 January (C) 6 January (D) 7 January [1]

4 Time in New Zealand is 2 hours ahead of Sydney. If it is 11:30 am in Sydney, what time is it in New Zealand?
(A) 1:30 pm (B) 9:30 pm (C) 1:30 am (D) 9:30 am [1]

5 How many hours are in a week?
(A) 84 (B) 120 (C) 144 (D) 168 [1]

6 How long is it from 9:35 pm Monday until 2:25 pm Tuesday?
(A) 16 h 10 min (B) 17 h 10 min (C) 16 h 50 min (D) 17 h 50 min [1]

7 What is the time 140 minutes after 6:20 am?
(A) 8 am (B) 8:53 am (C) 7:06 am (D) 8:40 am [1]

8 How many minutes are in $5\frac{3}{4}$ hours?
(A) 345 (B) 375 (C) 545 (D) 575 [1]

9 What is 7 h 25 min + 3 h 55 min?
(A) 11 h 20 min (B) 10 h 8 min (C) 10 h 48 min (D) 11 h 30 min [1]

10 To the nearest hour, how long would it take to travel 525 km at an average speed of 71 km/h?
(A) 6 h (B) 7 h (C) 8 h (D) 9 h [1]

11 A TV program starts at 1935 and finishes at 2110. How long is the program?
(A) 1 h 25 min (B) 1 h 35 min (C) 2 h 25 min (D) 2 h 35 min [1]

12 What is the date 3 weeks after 8 August?
(A) 23 August (B) 28 August (C) 29 August (D) 30 August [1]

13 What is the time on Tuesday $20\frac{1}{2}$ hours after 6:45 pm Monday?
(A) 1:35 pm (B) 2:15 pm (C) 2:35 pm (D) 3:15 pm [1]

14 Jed used his calculator when he wanted to divide 58 minutes by 8. His answer was 7.25. What is that?
(A) $7\frac{1}{4}$ seconds (B) 7 min 25 s (C) 7 min 15 s (D) none of these [1]

15 Sue arrives home at half past four in the afternoon. Her favourite program starts at 1950. How long does she have to wait?
(A) 1 h 20 min (B) 3 h 20 min (C) 3 h (D) 15 h [1]

PART B Write answers only in the space provided.

16 What is 3.5 minutes in minutes and seconds? ______________ 1

17 How many minutes are in $3\frac{3}{4}$ hours? ______________ 1

18 How long is it from 11:35 am until 2:25 pm? ______________ 1

19 What time is a quarter to eight at night in 24-hour time? ______________ 1

20 To the nearest hour, how long is 500 minutes? ______________ 1

21 What is the date 10 days after 26 August? ______________ 1

22 What is the time $5\frac{1}{4}$ hours after 6:50 am? ______________ 1

23 How many days are in winter (June, July and August)? ______________ 1

24 If it is 6 pm in Melbourne, what time is it in London which is 10 hours behind? ______________ 1

25 What is 37 minutes and 40 seconds plus 19 minutes and 40 seconds? ______________ 1

PART C Show all working.

26 Bec looks at this clock in the afternoon.

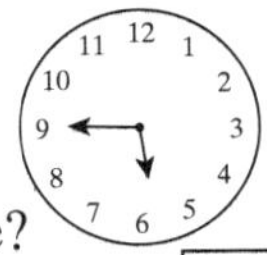

a What is the time in 24-hour time? 1 mark

____

b Bec's favourite program begins at 7:35 pm. How long does she have to wait? 1 mark

____

c The program runs for 2 hours and 17 minutes. What time will it finish? 1 mark

____

27 This is a timetable for an express train.

Bond	09:56
Redhill	10:19
Finch	10:42
Ashton	11:07
Melville	11:38
City	13:16

a At what 12-hour time does the train reach the city? 1 mark

____

b How long does the train take to travel from Bond to the city? 1 mark

____

c John gets to Bond station at twenty to ten. How long does he have before the train is due to depart? 1 mark

____

d If the train is running 15 minutes late when it leaves Melville, at what time will it leave Melville? 1 mark

____

e Meg wanted to be at Ashton station at least a quarter of an hour before the train is due to leave. What is the latest time Meg can arrive at the station? 1 mark

____

f Andy has a meeting in the city at 2:45 pm. How long does he have to wait if the train arrives in the city on time? 1 mark

____

g If Andy's meeting runs for $1\frac{1}{2}$ hours and it takes 20 minutes for Andy to return to the station, will he be able to catch the return train that leaves at 1635? Justify your answer. 2 marks

____

____

28 Susan drove 405 km and worked out her average speed was 75 km/h.

a For how long, in hours and minutes, was Susan driving? 2 marks

____

____

b Susan left home at 8:48 am and stopped for half an hour during the journey. What time did she arrive? 2 marks

____

____

Total marks /40

Time Calculator Allowed

LEVEL 2 TEST PART A

Time allowed: 40 minutes

1 What is the date two weeks after 22 October?
Ⓐ 1 November Ⓑ 2 November Ⓒ 5 November Ⓓ 6 November [1]

2 Time in Perth is an hour and a half behind Adelaide. What is the time in Perth if it is 4:05 pm in Adelaide?
Ⓐ 5:35 pm Ⓑ 2:55 pm Ⓒ 3:35 pm Ⓓ 2:35 pm [1]

3 How many hours is it from 8:30 am Monday to 5:30 pm Friday of the same week?
Ⓐ 57 Ⓑ 99 Ⓒ 103 Ⓓ 105 [1]

4 How long is 384 minutes in hours and minutes?
Ⓐ 6 h 24 min Ⓑ 6 h 14 min Ⓒ 6 h 4 min Ⓓ 6 h 40 min [1]

5 To the nearest minute, how long would it take to pump 3200 mL at 18 mL per second?
Ⓐ 1 min Ⓑ 2 min Ⓒ 3 min Ⓓ 4 min [1]

6 Mary used her calculator to divide 258 minutes by 40 and got 6.45. How long is that?
Ⓐ 6.45 s Ⓑ 6 min 45 s Ⓒ 6 min 27 s Ⓓ 6 min 23 s [1]

7 What is the time 210 minutes after 10:45 am?
Ⓐ 12:55 am Ⓑ 12:55 pm Ⓒ 2:15 pm Ⓓ 2:35 pm [1]

8 A movie starts at 1440 and runs for $2\frac{3}{4}$ hours. When does it finish?
Ⓐ 5:15 pm Ⓑ 5:25 pm Ⓒ 7:15 pm Ⓓ 7:25 pm [1]

9 What is the date 40 days after 25 April?
Ⓐ 3 June Ⓑ 4 June Ⓒ 5 June Ⓓ 6 June [1]

10 How many seconds are in 3 hours, 25 minutes and 45 seconds?
Ⓐ 15 000 Ⓑ 23 145 Ⓒ 11 745 Ⓓ 12 345 [1]

11 How many minutes are in 9.4 hours?
Ⓐ 564 Ⓑ 580 Ⓒ 554 Ⓓ 544 [1]

12 What is 15 h 36 min + 18 h 48 min?
Ⓐ 33 h 54 min Ⓑ 34 h 24 min Ⓒ 34 h 14 min Ⓓ 34 h 34 min [1]

13 A sports telecast begins at 1340 and finishes at 1625. How long is it on air?
Ⓐ 2 h 55 min Ⓑ 2 h 45 min Ⓒ 3 h 5 min Ⓓ 3 h 15 min [1]

14 What is the time on Saturday morning $28\frac{3}{4}$ hours after 10:45 pm Thursday?
Ⓐ 2:00 am Ⓑ 2:30 am Ⓒ 3:00 am Ⓓ 3:30 am [1]

15 How long is it from 2135 Wednesday to 1115 Thursday?
Ⓐ 13 h 40 min Ⓑ 13 h 20 min Ⓒ 14 h 40 min Ⓓ 10 h 20 min [1]

PART B Write answers only in the space provided.

16 What is 7.25 hours in hours and minutes? ______________ 1

17 How long is it from 10:45 am until 12:15 pm? ______________ 1

18 What is the time and day $20\frac{3}{4}$ hours after 5:30 pm Monday? ______________ 1

19 How many days is 888 hours? ______________ 1

20 What is the date 35 days after 15 November? ______________ 1

21 How long, in hours and minutes, is it from 1440 Tuesday to 0930 Wednesday? ______________ 1

22 Time in Sydney is 15 hours ahead of New York. If it is 7:30 am Saturday in Sydney, what time and day is it in New York? ______________ 1

23 What is the time $5\frac{1}{4}$ hours before 11 pm? ______________ 1

24 How many seconds are in $2\frac{1}{2}$ hours? ______________ 1

25 What is 19 h 35 min + 17 h 45 min in days, hours and minutes? ______________ 1

PART C Show all working.

26 On Friday, Tom worked for 3 h 45 min in the morning and 4 h 35 min in the afternoon.

a In total, how long did Tom work on Friday? ______________ 1 mark

b If Tom began work at 8:20 am and finished at 5:05 pm, for how long was he stopped for lunch? ______________ 2 marks

c Tom worked for exactly half the amount of time on Saturday than on Friday. If he began work at the same time and worked without a break, at what time did Tom finish on Saturday? ______________ 1 mark

27 Mia and Charlie are getting married at 3 pm on Saturday, 30 September.

a Mia needs to finalise plans for the reception 6 weeks before the wedding. What is the final date? ______________ 1 mark

b Mia and Charlie will return from their honeymoon on 13 October. What day of the week is that? ______________ 1 mark

c Mia's grandmother lives in a country that is 9 hours behind in time. At what time does the grandmother need to log on to watch the wedding? ______________ 1 mark

d On their wedding day, Mia and Charlie will have known each other for exactly 1000 days. About how long is that in years and months? ______________ 1 mark

28 This is the movie program for the Deluxe Cinema over a long weekend, showing the time each movie commences and the length of time it runs.

Movie	Sat.	Sun.	Mon.
The Menace (1 h 53 min)	6:30 pm	8:15 pm	5:55 pm
Flowering Peach (2 h 26 min)	8:40 pm	5:30 pm	8:10 pm
Messy Moments (2 h 5 min)	2:50 pm	12:45 pm	10:45 am
Steam-shovel Ted (1 h 25 min)	10:45 am	3:15 pm	4:10 pm
Woolly Wizard (1 h 45 min)	12:30 pm	10:45 am	2:05 pm

a What time does *Flowering Peach* start on Sunday, in 24-hour time? ______________ 1 mark

b Abigail is watching a movie at that cinema at 1:20 pm on Sunday. What movie is she watching? ______________ 1 mark

c What time will *Steam-shovel Ted* finish on Saturday? ______________ 1 mark

d What is the difference in running time between *Messy Moments* and *Woolly Wizard*? ______________ 1 mark

e How many minutes are there between the finish of *The Menace* and the start of *Flowering Peach* on Monday? ______________ 1 mark

f How much later does the final movie finish on Saturday than on Sunday? ______________ 2 marks

Total marks /40

LEVEL 3 TEST PART A

Time allowed: 40 minutes

1 How long is it from 8:40 am Friday until 7:20 pm Monday?
(A) 82 h 20 min (B) 82 h 40 min (C) 83 h 20 min (D) 83 h 40 min [1]

2 Sydney time is 10 hours ahead of London and Los Angeles time is 8 hours behind London. If it is 8 am Sunday in Sydney, what time is it in Los Angeles?
(A) 2 pm Sunday (B) 2 am Sunday (C) 2 pm Saturday (D) 2 am Saturday [1]

3 What is the date 6 weeks after 19 March?
(A) 30 April (B) 31 April (C) 1 May (D) 2 May [1]

4 If each cycle of a machine takes 23 minutes, how long will 24 cycles take?
(A) 9 h 2 min (B) 9 h 22 min (C) 9 h 20 min (D) 9 h 12 min [1]

5 A sports match overseas is shown on television at night from 2238 until 0618. How long is that?
(A) 7 h 20 min (B) 7 h 40 min (C) 8 h 20 min (D) 8 h 40 min [1]

6 19 h 45 min + 36 h 25 min is 2 days and ?
(A) 8 h 10 min (B) 7 h 42 min (C) 7 h 7 min (D) 8 h 20 min [1]

7 Eli divided 469 minutes by 35 and got 13.4. What is that?
(A) 13.4 s (B) 13 min 4 s (C) 13 min 24 s (D) 13 min 40 s [1]

8 How many days are there from, and including, 13 September until, and including, 17 October?
(A) 33 (B) 34 (C) 35 (D) 36 [1]

9 About how long is one million minutes?
(A) 9 months (B) $2\frac{1}{2}$ months (C) 1 year 9 months (D) 1 year 11 months [1]

10 A television program is scheduled to begin at 2125 and run for $3\frac{3}{4}$ hours. When does it finish?
(A) 0110 (B) 0010 (C) 0105 (D) 0100 [1]

11 What is the date 100 days after 14 February in a leap year?
(A) 23 May (B) 24 May (C) 25 May (D) 26 May [1]

12 How many hours are from 3:15 pm 2 July to 9:15 am 20 July?
(A) 414 (B) 450 (C) 426 (D) 438 [1]

13 What is the time 712 minutes after 0318?
(A) 1030 (B) 1410 (C) 1530 (D) 1510 [1]

14 What is the time and day $36\frac{1}{4}$ hours after 11:50 am Wednesday?
(A) 12:05 pm Thursday (B) 12:05 am Friday
(C) 12:05 pm Friday (D) 12:05 am Saturday [1]

15 To the nearest hour, how long would it take to pump 400 000 L at 5 L per second?
(A) 22 hours (B) 23 hours (C) 24 hours (D) 25 hours [1]

PART B Write answers only in the space provided.

16 How many minutes are in 2 weeks, 3 days and $5\frac{1}{2}$ hours? ________________ 1

17 What is 4.35 hours in hours and minutes? ________________ 1

18 To the nearest month, how long is 40 weeks? ________________ 1

19 What is the date 75 days after 18 July? ________________ 1

20 To the nearest minute, how long is 1000 seconds? ________________ 1

21 How long is it, in hours and minutes, from 1355 Monday until 1040 on Wednesday? ____________ 1

22 What is the time and date 38 hours after 11:15 pm on 2 April? ________________ 1

23 If a movie that runs for 2 h 48 min begins at 10:35 am, when will it finish? ________________ 1

24 What is 3265 minutes in days, hour and minutes? ________________ 1

25 When it is 9:15 am in Adelaide it is 7:45 am in Perth. When it is 9:15 pm in Perth what is the time in Adelaide? ________________ 1

PART C Show all working.

26 Travis has a meeting at 1535. It is now twenty past two in the afternoon.

a How long does Travis have before the meeting? [1 mark]

____

b The meeting lasts for 1.25 hours. At what 12-hour time does the meeting finish? [2 marks]

____

____

27 In a medley relay, the backstroke swimmer swam 59.19 seconds, the breaststroke was completed in 1 minute and 6.68 seconds, the butterfly leg in 57.05 seconds and the freestyle in 53.04 seconds. The world record was 3 minutes 52.05 seconds. Did the team break the record? If so, by how many seconds or, if not, how many seconds faster did they need to swim? [2 marks]

____

____

28 a A plane leaves Sydney at 11:10 am and flies to Hobart arriving at five past one in the afternoon. How long was the flight? [1 mark]

____

b If the return flight leaves Hobart at 1635 and takes the same amount of time, at what 12-hour time will it land in Sydney? [1 mark]

____

c Another plane leaves Sydney at 10:50 am Tuesday and flies to Johannesburg in South Africa. The flight takes 14 hours and 20 minutes. What is the time and day in Sydney when the plane reaches Johannesburg? [1 mark]

____

d Sydney is 8 hours ahead of Johannesburg time. What is the time and day in Johannesburg when the plane lands? [1 mark]

____

e The return flight leaves Johannesburg at 1910 on Wednesday. Assuming that the flight takes the same amount of time, what would be the time and day in Sydney when the plane lands? [2 marks]

____

____

29 a At 11:30 am on Wednesday water began to flow into a large empty tank. At 2:10 pm on Friday the tank reached its full capacity of 45 600 L. How many litres were flowing each minute? [2 marks]

____

____

b When would the tank have held 25 000 L? Give the answer to the nearest quarter hour. [2 marks]

____

____

Total marks /40

Mixed Measurement Calculator Allowed

LEVEL 1 TEST — PART A

Time allowed: 40 minutes

1 A square paddock of side length 300 m would have an area of:
(A) 3 ha (B) 9 ha (C) 30 ha (D) 90 ha — 1

2 2 m^3 is the same as:
(A) 200 cm^3 (B) 20 000 cm^3 (C) 200 000 cm^3 (D) 2 000 000 cm^3 — 1

3 The triangular face of a triangular prism has area 60 cm^2. If the height of the prism is 8 cm, what is the volume?
(A) 240 cm^3 (B) 160 cm^3 (C) 120 cm^3 (D) 480 cm^3 — 1

4 What is the time $3\frac{3}{4}$ hours after 11:30 am?
(A) 3:15 pm (B) 3:05 pm (C) 2:45 pm (D) 3:25 pm — 1

5 9 ML =
(A) 9000 L (B) 900 000 L (C) 9 000 000 L (D) 9 000 000 000 L — 1

6 What is 2.2 hours in hours and minutes?
(A) 2 h 2 min (B) 2 h 12 min (C) 2 h 20 min (D) 2 h 22 min — 1

7 Which is correct for this triangle?

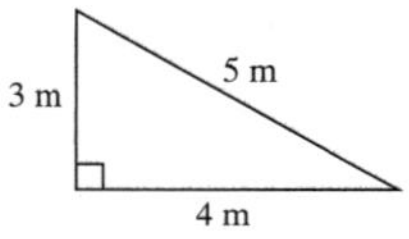

(A) perimeter = 6 m, area = 6 m^2 (B) perimeter = 6 m, area = 12 m^2
(C) perimeter = 12 m, area = 6 m^2 (D) perimeter = 12 m, area = 12 m^2 — 1

8 What is the area of a kite if one diagonal is 26 cm long and the other 15 cm long?
(A) 195 cm^2 (B) 97.5 cm^2 (C) 130 cm^2 (D) 390 cm^2 — 1

9 How long is 500 hours?
(A) 20 d 8 h (B) 20 d 20 h (C) 41 d 6 h (D) 41 d 8 h — 1

10 What is the capacity of a rectangular prism 10 cm long, 3 cm wide and 5 cm high?
(A) 1.5 mL (B) 15 mL (C) 150 mL (D) 1500 mL — 1

11 What is the circumference of a circle, to the nearest centimetre, if the diameter is 14 cm?
(A) 22 cm (B) 44 cm (C) 88 cm (D) 1.54 m — 1

12 What is the area of this trapezium?

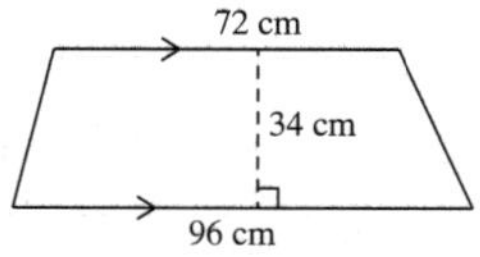

(A) 5712 cm^2 (B) 5088 cm^2 (C) 4680 cm^2 (D) 2856 cm^2 — 1

13 What is the exact area of a circle of radius 6 cm?
(A) 6π cm^2 (B) 9π cm^2 (C) 12π cm^2 (D) 36π cm^2 — 1

14 What is the area of a rhombus if one diagonal is 10 cm and the other 8 cm long?
(A) 160 cm^2 (B) 80 cm^2 (C) 40 cm^2 (D) 20 cm^2 — 1

15 Which is closest to the volume of a cylinder of radius 18 cm and height 13 cm?
(A) 13 230 cm^3 (B) 3310 cm^3 (C) 9560 cm^3 (D) 52 930 cm^3 — 1

PART B Write answers only in the space provided.

16 How many litres are in 320 kL? ____________ 1

17 What is the area of a parallelogram of base length 9 cm and perpendicular height 6 cm? ____________ 1

18 What is the perimeter of a rhombus of side length 8 cm? ____________ 1

19 How many minutes are in $7\frac{3}{4}$ hours? ____________ 1

20 A container has volume 750 cm^3. What is its capacity in millilitres? ____________ 1

21 What is the circumference, in terms of π, of a circle of diameter 5 cm? ____________ 1

22 What is the volume of a prism if the area of the cross-section is 40 cm^2 and the perpendicular height is 7 cm? ____________ 1

23 What is the date 21 days after 25 July? ____________ 1

24 What is the area of a circle, to the nearest square centimetre, if the radius is 7 cm? ____________ 1

25 How many cubic centimetres are in 8 m^3? ____________ 1

PART C Show all working.

26 What is the area of this paddock, in hectares? [2 marks]

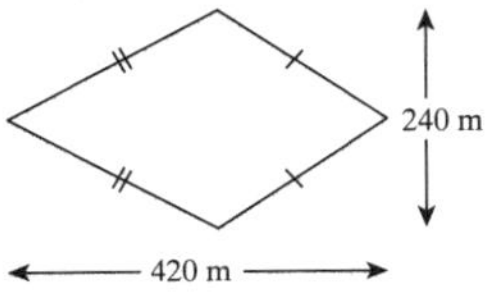

____

____

27 A movie begins at a quarter to eight at night and runs for 2 hours and 25 minutes. At what 24-hour time will it finish? [2 marks]

____

____

28 This triangle has perimeter 2.6 m.

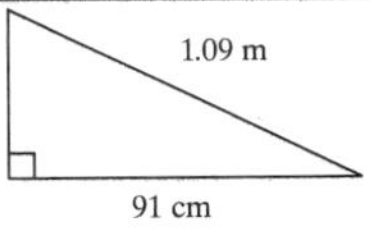

a What is the length of the third side in centimetres? [1 mark]

____

b What is the area of the triangle in square metres? [1 mark]

____

c If the triangle forms the cross-section of a prism of height 75 cm, what is the volume in cubic metres, correct to three decimal places? [2 marks]

____

____

29 A cylindrical tank has radius 1.2 m and height 2.2 m.

a What is the capacity of the tank in litres, to the nearest 100 L? [2 marks]

____

____

b A tap was left open for 2 hours and 40 minutes. In that time 2000 L flowed out of the tank. How many litres is that per minute? [2 marks]

____

____

30 A window is in the shape of a semicircle above a square.

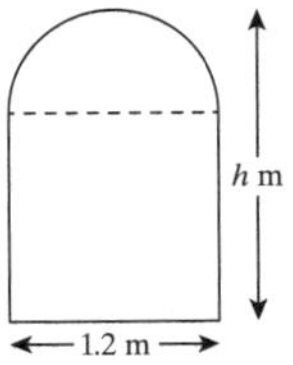

a What is the height, h m, of the window? [1 mark]

____

b What is the approximate area of the window? [2 marks]

____

____

____

____

Total marks /40

LEVEL 2 TEST PART A

Time allowed: 40 minutes

1 What would be the exact area of a circle of diameter 4 cm?

(A) 2π cm² (B) 4π cm² (C) 8π cm² (D) 16π cm² 1

2 56 cm³ =

(A) 560 mm³ (B) 5600 mm³ (C) 56 000 mm³ (D) 560 000 mm³ 1

3 What is the capacity of a rectangular prism 5 cm long, 4 cm wide and 6 cm high?

(A) 120 mL B 1.2 L (C) 12 L (D) 120 L 1

4 What is the date 60 days after 16 August?

(A) 14 October (B) 15 October (C) 16 October (D) 17 October 1

5 What is the circumference, to the nearest centimetre, of a circle of radius 47 cm?

(A) 1.48 m (B) 84 cm (C) 2.95 m (D) 5.91 m 1

6 What is the volume of this prism?

(A) 120 m³ (B) 180 m³ (C) 360 m³ (D) 1020 m³ 1

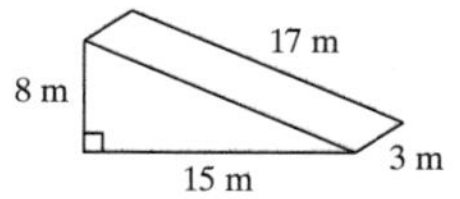

7 What is the area of a kite if one diagonal is 88 cm and the other 1.25 m?

(A) 5500 cm² (B) 11 000 cm² (B) 55 000 cm² (D) 110 000 cm² 1

8 A semicircle has area 50π cm². What is the diameter?

(A) 50 cm (B) 20 cm (C) 10 cm (D) 5 cm 1

9 What is the area of this trapezium?

(A) 800 cm² (B) 720 cm² (C) 640 cm² (D) 696 cm² 1

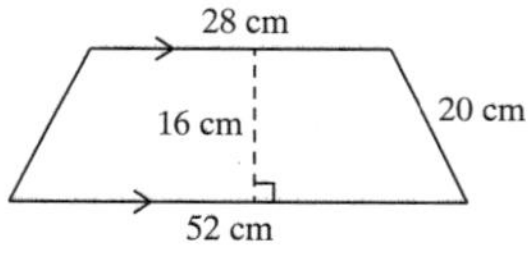

10 What is the exact volume of a cylinder of diameter 6 cm and height 2 cm?

(A) 18π cm³ (B) 36π cm³ (C) 144π cm³ (D) 72π cm³ 1

11 To the nearest minute, how long would it take to pump 20 L at 5 mL per second?

(A) 16 min (B) 17 min (C) 66 min (D) 67 min 1

12 How many square metres are in 200 cm²?

(A) 2 (B) 0.2 (C) 0.02 (D) 0.002 1

13 5 kL and 20 mL is:

(A) 50 000.2 L (B) 5000.2 L (C) 50 000.02 L (D) 5000.02 L 1

14 What is the area of a right-angled triangle with sides 48 cm, 55 cm and 73 cm?

(A) 1320 cm² (B) 1752 cm² (C) 1760 cm² (D) 2007.5 cm² 1

15 Which is closest to the area of a rhombus if one diagonal is 17 m and the other 13 m?

(A) 166 m² (B) 55 m² (C) 220 m² (D) 110 m² 1

Mixed Measurement Calculator Allowed **(Level 2 Test)**

PART B Write answers only in the space provided.

16 What is the area of a square if the perimeter is 20 cm? ______________ [1]

17 What is the area, in terms of π, of a circle of diameter 22 cm? ______________ [1]

18 What is the capacity, in litres, of a rectangular prism 3 m long, 2 m wide and 1 m high? ____________ [1]

19 How many hectares are in a square kilometre? ______________ [1]

20 How many minutes are in 4.4 hours? ______________ [1]

21 What is the area of a rhombus, in square metres, if one diagonal is 1.4 m and the other 80 cm? ______________ [1]

22 What is the volume of a triangular prism if the area of the cross-section is 47 m^2 and the perpendicular height 15 m? ______________ [1]

23 What is the circumference, to the nearest centimetre, of a circle of radius 28 cm? ______________ [1]

24 How many cubic centimetres are in 0.2 m^3? ______________ [1]

25 How long is it from 1145 Monday to 1925 Tuesday? ______________ [1]

PART C Show all working.

26 A driveway is formed by one part of a rectangle being replaced by a quadrant as shown in the diagram.

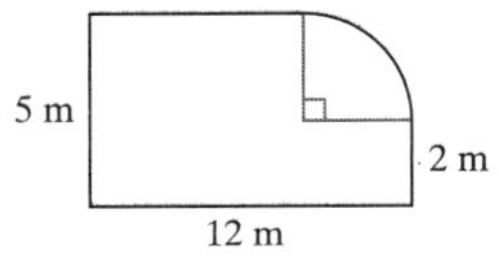

a What is the radius of the quadrant? [1 mark]

____

b What is the total area of the driveway? Give the answer to two decimal places. [2 marks]

____

____

c If the driveway is concreted to a depth of 15 cm, what amount of concrete will be needed? (Give the answer in cubic metres to one decimal place.) [2 marks]

____

____

27

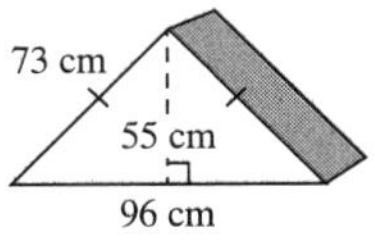

a What is the area of the triangular cross-section of this prism in square metres? [1 mark]

____

b If the volume of the prism is 1.32 m^3, what is the area of the shaded face? [2 marks]

____

____

28 This park is formed by a trapezium and kite.

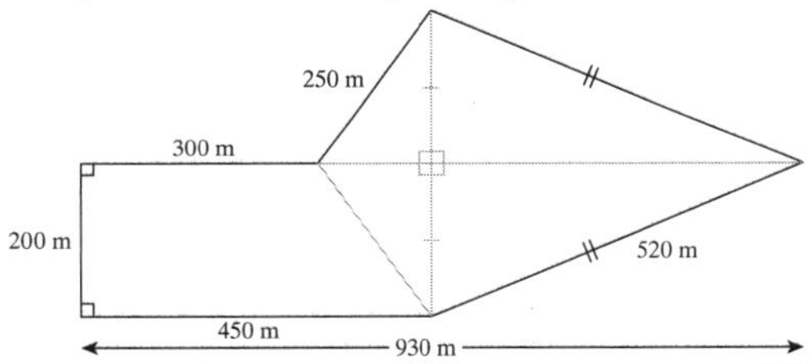

a What is the area of the trapezium in square metres? [1 mark]

____

b What is the area of the kite in square metres? [2 marks]

____

____

c What is the total area of the park in hectares? [1 mark]

____

d What is the perimeter of the park in kilometres? [1 mark]

____

e Corinne set off at 6:48 am and walked around the entire perimeter of the park, returning to her starting point at 7:16 am. What was her average speed in kilometres per hour? [2 marks]

____

____

Total marks /40

Mixed Measurement Calculator Allowed

LEVEL 3 TEST PART A

Time allowed: 40 minutes

1 Which is closest to the capacity of a rectangular prism 5.4 m long, 2.7 m wide and 2.8 m high?
Ⓐ 40 L Ⓑ 400 L Ⓒ 4 kL Ⓓ 40 kL 1

2 40 700 cm^3 =
Ⓐ 407 m^3 Ⓑ 4.07 m^3 Ⓒ 0.0407 m^3 Ⓓ 0.004 07 m^3 1

3 Which is closest to the area of a rhombus if one diagonal is 1.3 m long and the other 94 cm?
Ⓐ 1200 cm^2 Ⓑ 600 cm^2 Ⓒ 0.6 m^2 D 1.2 m^2 1

4 How many square kilometres is 257 ha?
Ⓐ 25.7 Ⓑ 2.57 Ⓒ 0.257 Ⓓ 0.0257 1

5 What is the area of this quadrilateral?
Ⓐ 277.5 cm^2 Ⓑ 330 cm^2 Ⓒ 555 cm^2 Ⓓ 528 cm^2 1

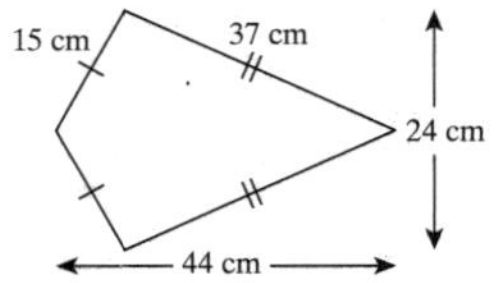

6 What is 3 hours and 37 minutes plus 4.3 hours?
Ⓐ 8 h 7 min Ⓑ 7 h 40 min Ⓒ 8 h 42 min Ⓓ 7 h 55 min 1

7 Which is closest to the area of a circle of diameter 13.3 m?
Ⓐ 50 m^2 Ⓑ 140 m^2 Ⓒ 280 m^2 Ⓓ 560 m^2 1

8 A triangle has base length 12 cm and perpendicular height 5 cm. It forms the cross-section of a prism of height 7 cm. What is the volume of the prism?
Ⓐ 140 cm^3 Ⓑ 210 cm^3 Ⓒ 240 cm^3 Ⓓ 420 cm^3 1

9 What is the perimeter, to the nearest centimetre, of a semicircle of radius 17 cm?
Ⓐ 107 cm Ⓑ 124 cm Ⓒ 87 cm Ⓓ 141 cm 1

10 What is the capacity, to the nearest 1000 L, of a cylinder of diameter 2.4 m and height 1.8 m?
Ⓐ 3000 L Ⓑ 8000 L Ⓒ 33 000 L Ⓓ 6000 L 1

11 What is the radius, to the nearest centimetre, of a circle with circumference 3.1 m?
Ⓐ 49 cm Ⓑ 50 cm Ⓒ 98 cm Ⓓ 99 cm 1

12 37 200 kL + 0.8 ML =
Ⓐ 38 ML Ⓑ 45.2 ML Ⓒ 837 200 kL Ⓓ 117 200 kL 1

13 The area of this quadrilateral is closest to:
Ⓐ 0.6 m^2 Ⓑ 0.9 m^2 Ⓒ 1.1 m^2 Ⓓ 1.3 m^2 1

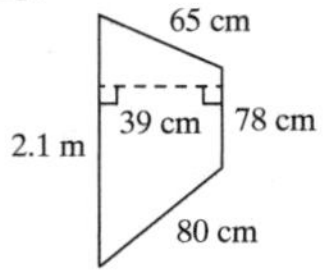

14 A sporting event begins on television at 2235 and runs for $7\frac{3}{4}$ hours. When will it finish?
Ⓐ 5:10 am Ⓑ 6:10 am Ⓒ 5:20 am Ⓓ 6:20 am 1

15 If 19 June is a Monday, what day of the week will 3 October be?
Ⓐ Sunday Ⓑ Monday Ⓒ Tuesday Ⓓ Wednesday 1

PART B Write answers only in the space provided.

16 What is the area, to the nearest square centimetre, of a quadrant of a circle of radius 34 cm? ______ [1]

17 How many litres are in 0.08 ML? ______ [1]

18 What is the area in hectares of a square of side 2 km? ______ [1]

19 A right-angled triangle forms the cross-section of a prism of height 5 cm. If the sides of the triangle are 6 cm, 8 cm and 10 cm, what is the volume of the triangular prism? ______ [1]

20 What is the area of a trapezium, in square metres, if the parallel sides are 37 cm and 58 cm long and the perpendicular distance between them is 42 cm? ______ [1]

21 A train leaves Penton at 10:46 am to travel to Upton. If the trip takes 3 hours and 18 minutes, at what time will the train arrive at Upton? ______ [1]

22 What is the capacity, in litres, of a prism with cross-sectional area 190 cm^2 and height 16 cm? ______ [1]

23 The circumference of a circle is 81.7 cm. What is its approximate radius? ______ [1]

24 How many square centimetres are in 0.09 m^2? ______ [1]

25 What is the volume of a cube if the area of the cross-section is 169 cm^2? ______ [1]

PART C Show all working.

26 The diagram shows a top view of a swimming pool surrounded by an area to be paved.

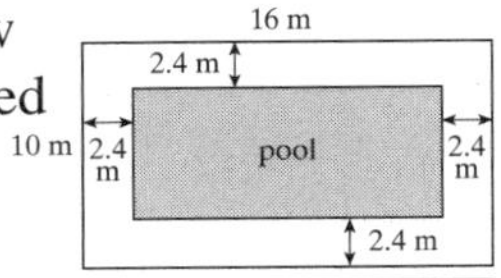

a Find the cost of a fence around the outside of the whole area at $78 per metre. [1 mark]

______

b If the pool is filled to a constant depth of 1.5 m, how many litres of water will it hold when full? [3 marks]

______

c If the pool is filled from a hose with water flowing at the rate of 15 L every minute, how long in days, hours and minutes will it take to fill the pool? [2 marks]

______

d Square pavers of side length 400 mm are to surround the pool. The pavers cost $9.00 each. What is the total cost of the pavers? [2 marks]

______

27 One diagonal of a rhombus is 1.6 m and the other is 78 cm.

a What is the area of the rhombus in square metres? [1 mark]

______

b If the perimeter of the rhombus is 3.56 m, what is the perpendicular height, to the nearest centimetre? [2 marks]

______

28 This shape is made up of a large semicircle less two smaller, equal-sized semicircles.

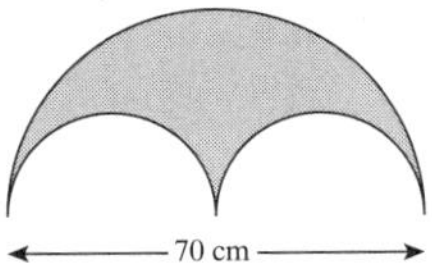

a Find the area of the shape to the nearest square centimetre. [2 marks]

______

b What is the exact perimeter of the shape? [2 marks]

______

Total marks /40

Pythagoras' Theorem Calculator Allowed

PRETEST

Time allowed: 15 minutes **Total marks 15**

1 $35^2 =$

Ⓐ 70 Ⓑ 925 Ⓒ 1225 Ⓓ 42 875 — 1

2 $\sqrt{531\,441} =$

Ⓐ 21 Ⓑ 81 Ⓒ 731 Ⓓ 729 — 1

3 $2^2 + 3^2 =$

Ⓐ 10 Ⓑ 13 Ⓒ 25 Ⓓ 49 — 1

4 $315^2 - 247^2 =$

Ⓐ 676 Ⓑ 4624 Ⓒ 24 149 Ⓓ 38 216 — 1

5 What is $\sqrt{39}$ correct to two decimal places?

Ⓐ 3.39 Ⓑ 6.24 Ⓒ 6.25 Ⓓ 15.21 — 1

6 $39^2 + 80^2 =$

Ⓐ 89^2 Ⓑ 91^2 Ⓒ 99^2 Ⓓ 101^2 — 1

7 $73^2 - 55^2 =$

Ⓐ 32^2 Ⓑ 38^2 Ⓒ 42^2 Ⓓ 48^2 — 1

8 Which angle of this quadrilateral is a right angle?

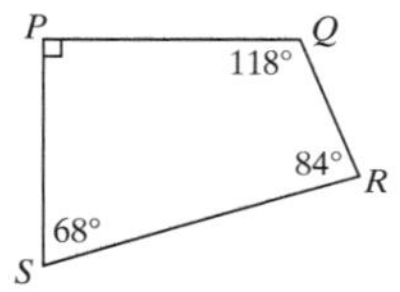

Ⓐ $\angle PQR$ Ⓑ $\angle QRS$ Ⓒ $\angle RSP$ Ⓓ $\angle SPQ$ — 1

9 $\sqrt{576 + 49} =$

Ⓐ 25 Ⓑ 31 Ⓒ 56 Ⓓ 73 — 1

10 $\sqrt{676} - \sqrt{100} =$

Ⓐ 4 Ⓑ 14 Ⓒ 16 Ⓓ 24 — 1

11 Which side is opposite the right angle?

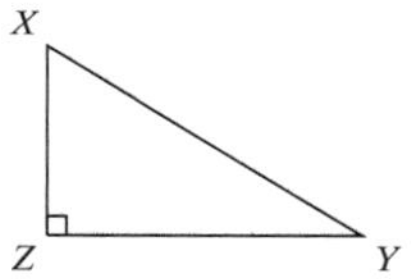

Ⓐ XZ Ⓑ YZ Ⓒ XY Ⓓ XZY — 1

12 What is the perimeter of this triangle?

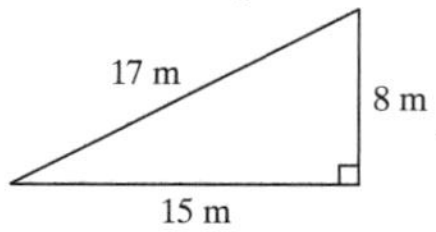

Ⓐ 40 m Ⓑ 50 m Ⓒ 60 m Ⓓ 70 m — 1

13 What is the area of the triangle in question 12?

Ⓐ 40 m² Ⓑ 60 m² Ⓓ 68 m² Ⓓ 127.5 m² — 1

14 $\sqrt{3^2 + 4^2} =$

Ⓐ 5 Ⓑ 7 Ⓒ 13 Ⓓ 19 — 1

15 $\sqrt{10^2 - 6^2} =$

Ⓐ 6 Ⓑ 8 Ⓒ 4 Ⓓ –26 — 1

Total marks /15

Pythagoras' Theorem Calculator Allowed

LEVEL 1 TEST PART A

Time allowed: 40 minutes

1 What is the length of the hypotenuse of this triangle?

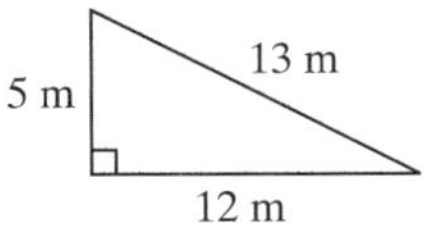

(A) 5 m (B) 12 m (C) 13 m (D) 30 m 1

2 Which is a rational number?

(A) $\sqrt{5}$ (B) $\sqrt{4}$ (C) $\sqrt{3}$ (D) $\sqrt{2}$ 1

3 What is the missing word in the following? 'In any right-angled triangle, the square on the hypotenuse is equal to the ________ of the squares on the other two sides.'

(A) sum (B) difference (C) product (D) quotient 1

4 What is $\sqrt{19}$ correct to one decimal place?

(A) 4.3 (B) 4.4 (C) 4.5 (D) 4.6 1

5 $15^2 + 20^2 =$

(A) 25^2 (B) 30^2 (C) 35^2 (D) 40^2 1

6 What is the value of x?

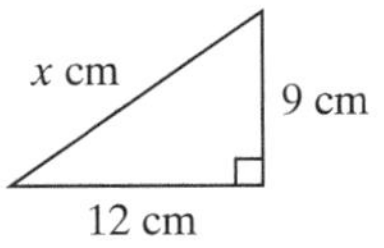

(A) 13 (B) 14 (C) 15 (D) 16 1

7 $\sqrt{260^2 + 651^2} =$

(A) 701 (B) 711 (C) 901 (D) 911 1

8 $37^2 - 35^2 =$

(A) 4 (B) 36 (C) 64 (D) 144 1

9 What is the value of h?

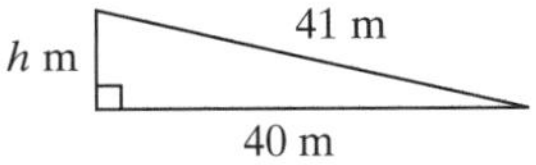

(A) 5 (B) 7 (C) 9 (D) 11 1

10 $\sqrt{653^2 - 315^2} =$

(A) 338 (B) 314 (C) 584 (D) 572 1

11 Which triangle must be right-angled?

(A) 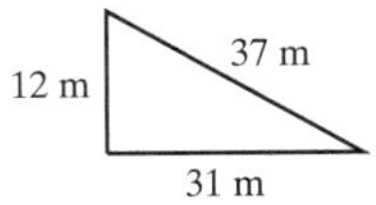

(B) 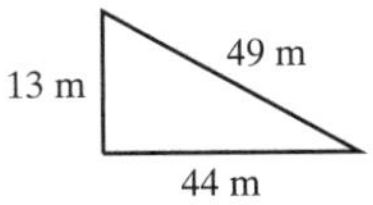

(C) 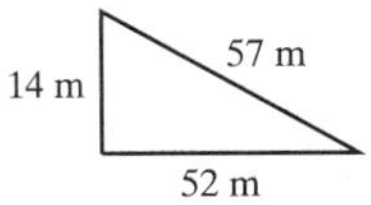

(D) 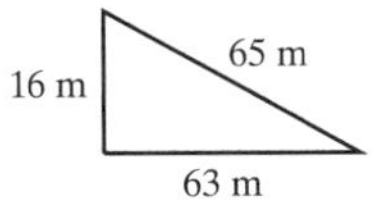

1

12 Which is a Pythagorean triad?

(A) 15, 112, 113 (B) 17, 63, 68 (C) 19, 44, 47 (D) 23, 28, 35 1

13 What would be the length of the hypotenuse of a right-angled triangle if the lengths of the other two sides are 32 cm and 60 cm?

(A) 51 cm (B) 64 cm (C) 68 cm (D) 72 cm 1

14 What is the exact value of x?

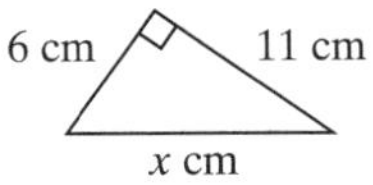

(A) $\sqrt{289}$ (B) $\sqrt{157}$ (C) $\sqrt{85}$ (D) $\sqrt{25}$ 1

15 Which triangle is **not** right-angled?

(A) 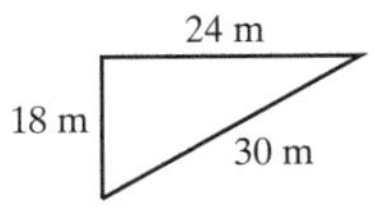

(B) 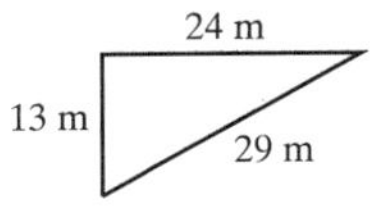

(C) 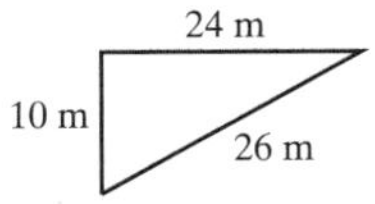

(D) 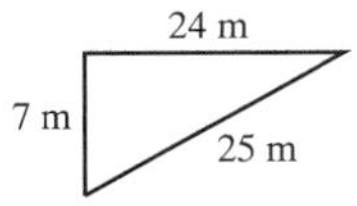

1

PART B Write answers only in the space provided.

16 $\sqrt{385641} =$ ____________ 1

17 What is $\sqrt{35}$ to two decimal places? ____________ 1

18 Is $\sqrt{16}$ rational or irrational? ____________ 1

19 What is the (positive) value of x if $x^2 = 39^2 + 80^2$? ____________ 1

20 $51^2 + 140^2 = 149^2$ True or false? ____________ 1

21 28, 195 and 197 form a Pythagorean triad. True or false? ____________ 1

22 What is the length of the hypotenuse of this right-angled triangle? ____________ 1

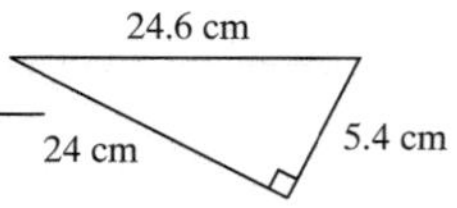

23 What is the (positive) value of h if $146 + h^2 = 267$? ____________ 1

24 What is the value of x? ____________ 1

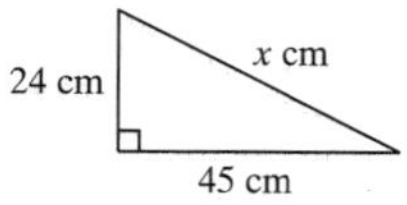

25 What is the value of h? ____________ 1

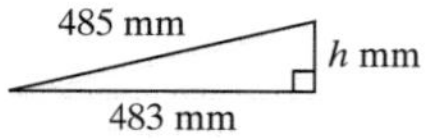

PART C Show all working.

26 Find the length of the unknown side of each triangle. Give the answers to the nearest centimetre.

a 2 marks

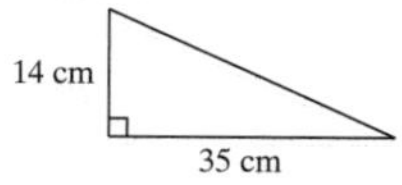

____

____

____

b 2 marks

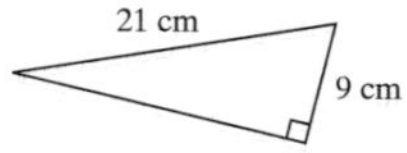

____

____

____

27 What is the distance across this lake? 2 marks

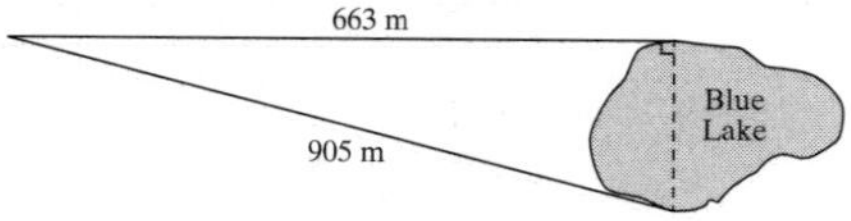

____

____

____

28 A gate is 14.5 m wide and 2.4 m high. How long, to the nearest centimetre, is the diagonal brace? 2 marks

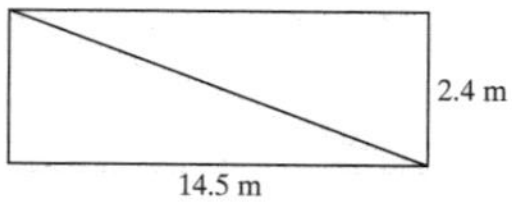

____

____

____

29 A ladder is 3.6 m long. It leans against a wall with its foot 1.4 m from the wall.

a Show this information on a diagram. 1 mark

b How far up the wall does the ladder reach? Give the answer in metres to one decimal place. 2 marks

____

____

____

30 a Find the length of AC in centimetres. 2 marks

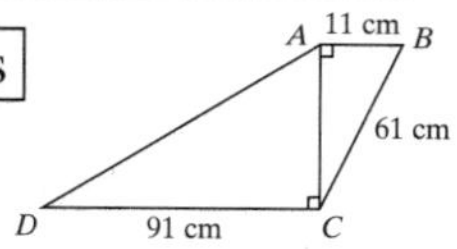

____

____

____

b Find the length of AD in metres. 2 marks

____

____

____

Total marks /40

Pythagoras' Theorem Calculator Allowed

LEVEL 2 TEST PART A

Time allowed: 40 minutes

1 The square of 64 is:

(A) 8 (B) 32 (C) 128 (D) 4096 [1]

2 If $x^2 = 196$ then x could equal:

(A) 14 (B) 16 (C) 24 (D) 36 [1]

3 $\sqrt{9048}$ is closest to:

(A) 94 (B) 95 (C) 96 (D) 97 [1]

4 What is the hypotenuse of this triangle?

(A) *PQ* (B) *PR* (C) *QR* (D) none of these [1]

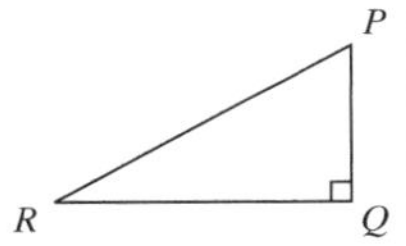

5 Which is **not** rational?

(A) $\sqrt{1369}$ (B) $\sqrt{961}$ (C) $\sqrt{1936}$ (D) $\sqrt{6139}$ [1]

6 The square root of seven hundred and twenty-nine is:

(A) 27 (B) 9 (C) 55 (D) 531 441 [1]

7 What is the length of the hypotenuse of this triangle?

(A) 0.6 m (B) 0.91 m (C) 1.09 m (D) 2.6 m [1]

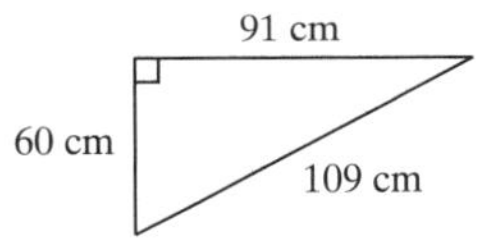

8 $\sqrt{5^2 + 12^2} =$

(A) 13 (B) 15 (C) 17 (D) 19 [1]

9 Which is correct? The hypotenuse of a right-angled triangle is:

I the longest side II the side opposite the right angle

(A) I only (B) II only (C) both I and II (D) neither I nor II [1]

10 $\sqrt{101^2 - 99^2} =$

(A) 2 (B) 4 (C) 12 (D) 20 [1]

11 A right-angled triangle has sides of 3.835 m, 372 mm and 385.3 cm. How long is the hypotenuse?

(A) 3.835 m (B) 372 mm (C) 385.3 cm (D) not possible to tell [1]

12 If $h^2 = 1322$, and h is positive, then h is closest to:

(A) 37 (B) 36 (C) 35 (D) 34 [1]

13 What is the value of x?

(A) 79.9 (B) 61.5 (C) 55.9 (D) 56.5 [1]

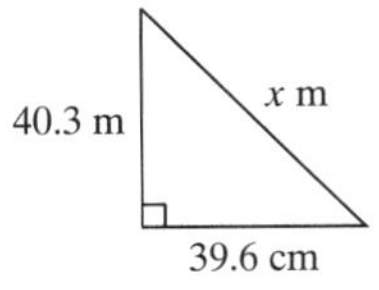

14 Which is a Pythagorean triad?

(A) 64, 167, 185 (B) 74, 177, 195 (C) 84, 187, 205 (D) 94, 197, 215 [1]

15 What is the value of h?

(A) 5.32 (B) 6.91 (C) 9.90 (D) 20.43 [1]

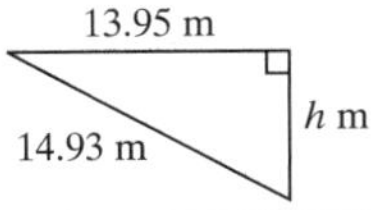

PART B Write answers only in the space provided.

16 What is $\sqrt{8649}$? ____________ [1]

17 Is $\sqrt{416}$ rational or irrational? ____________ [1]

18 If $x^2 = 7569$ and x is a positive number what is its value? ____________ [1]

19 115, 252 and 277 form a Pythagorean triad. True or false? ____________ [1]

20 $119^2 - 105^2 = 56^2$. True or false? ____________ [1]

21 What is the value of x? ____________ [1]

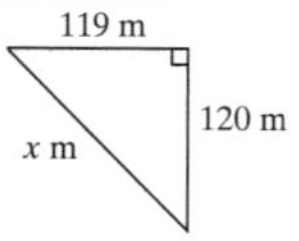

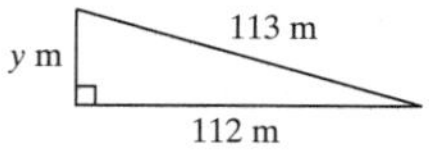

22 What is the value of y? ____________ [1]

23 What is the length of the hypotenuse of a right-angled triangle if the other sides measure 16 cm and 30 cm? ____________ [1]

24 Find $\sqrt{25^2 - 24^2}$. ____________ [1]

25 The hypotenuse of a right-angled triangle is 91 cm and another side is 35 cm. What is the length of the remaining side? ____________ [1]

PART C Show all working.

26 Find the length of the unknown side of each triangle. Give each answer correct to one decimal place.

a [2 marks]

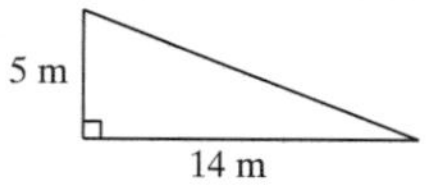

b [2 marks]

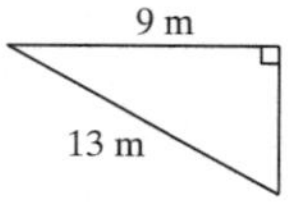

c [2 marks]

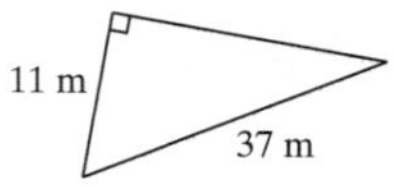

27 Find the area of this triangle. [2 marks]

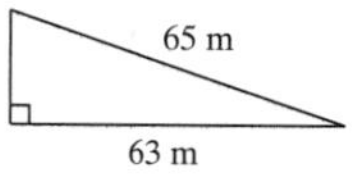

28 a Find the length of the hypotenuse of ΔABC. [1 mark]

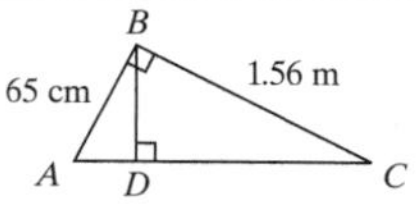

b Find the area, in square centimetres, of ΔABC. [1 mark]

c Find the length of BD. [1 mark]

d Find the length of AD. [1 mark]

29 The diagonals of a rhombus bisect each other at right angles. Find the perimeter of a rhombus that has one diagonal 48 cm long and the other 64 cm long. [3 marks]

Total marks /40

Pythagoras' Theorem Calculator Allowed

LEVEL 3 TEST **PART A**

Time allowed: 40 minutes

1 What is the length of the hypotenuse of this isosceles right-angled triangle?

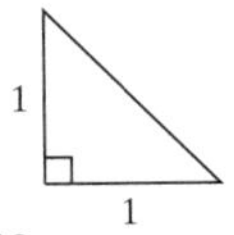

(A) 1 (B) $\sqrt{2}$ (C) $\sqrt{3}$ (D) 2 — 1

2 If *a*, *b* and *c* form a Pythagorean triad which of these must also form a Pythagorean triad?

(A) $x + a, x + b, x + c$ (B) $a - x, b - x, c - x$ (C) ax, bx, cx (D) $\frac{x}{a}, \frac{x}{b}, \frac{x}{c}$ — 1

3 Which is closest to the length of the unknown side of this triangle?

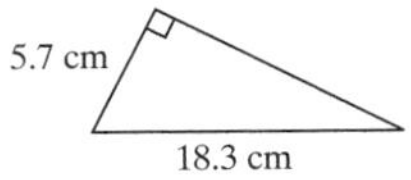

(A) 17.4 cm (B) 18.1 cm (C) 16.6 cm (D) 19.2 cm — 1

4 Which triangle does **not** have a hypotenuse exactly 65 cm long?

(A) 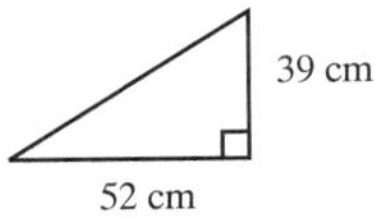

(B) 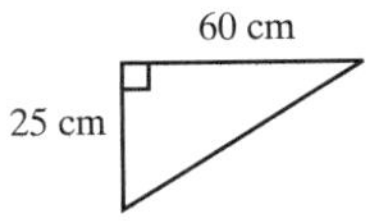

(C) 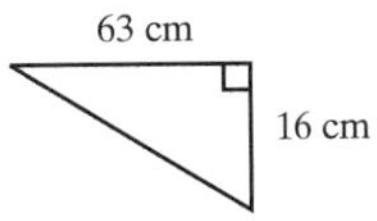

(D) 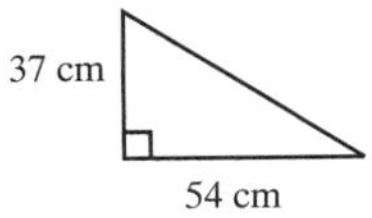

1

5 Which numbers form a Pythagorean triad?

(A) 86, 237, 255 (B) 96, 247, 265 (C) 106, 287, 305 (D) 116, 277, 295 — 1

6 Which is **not** a right-angled triangle? (All measurements are in centimetres.)

(A) 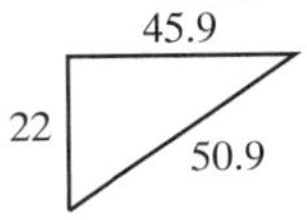

(B) 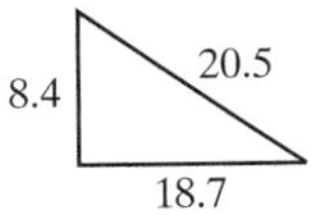

(C) 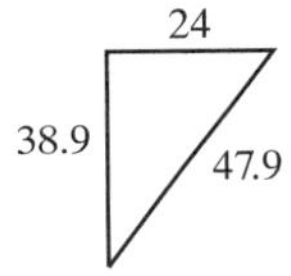

(D) 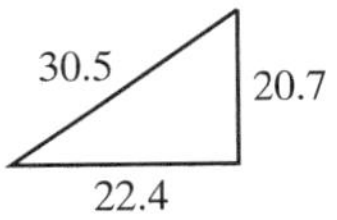

1

7 If $x^2 = 442^2 + 759^2 - 342^2$ then $x =$

(A) 809 (B) 849 (C) 859 (D) 869 — 1

8 What is the perimeter of this triangle?

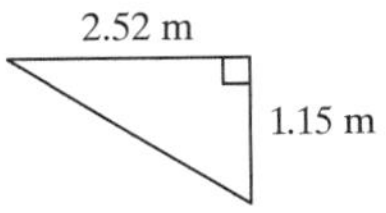

(A) 6.3 m (B) 6.44 m (C) 6.7 m (D) 6.94 m — 1

9 What is the area of this triangle?

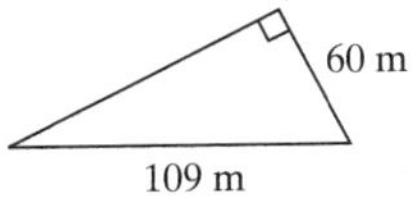

(A) 2730 m² (B) 3270 m² (C) 4560 m² (D) 5460 m² — 1

10 Which of these is irrational?

(A) $\sqrt{3969}$ (B) $\sqrt{4489}$ (C) $\sqrt{4909}$ (D) $\sqrt{5329}$ — 1

11 What number will, together with 135 and 377, form a Pythagorean triad?

(A) 266 (B) 352 (C) 396 (D) 400 — 1

12 What is the length of the longest side of a right-angled triangle if one of the other sides is $\frac{3}{4}$ m long and the third is one metre long?

(A) $1\frac{1}{4}$ m (B) $1\frac{1}{3}$ m (C) $1\frac{2}{3}$ m (D) $1\frac{1}{2}$ m — 1

13 Which is **not** a Pythagorean triad?

(A) 28, 195, 197 (B) 32, 255, 257 (C) 36, 323, 325 (D) 40, 389, 391 — 1

14 The square of a number will never end in which of these digits?

(A) 4 (B) 6 (C) 8 (D) 9 — 1

15 What is the value of *x*?

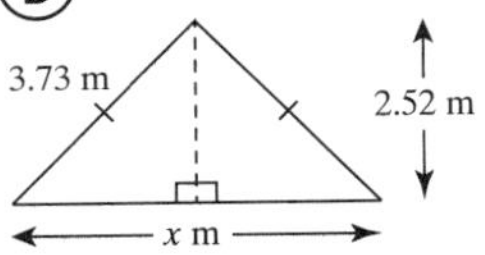

(A) 1.5 (B) 2.25 (C) 2.75 (D) 5.5 — 1

PART B Write answers only in the space provided.

16 Is $\sqrt{859}$ rational or irrational? ______________ [1]

17 $\sqrt{12^2 + 35^2}$ = ______________ [1]

18 What is $\sqrt{94}$ correct to two decimal places? ______________ [1]

19 What is the value of x? ______________ [1]

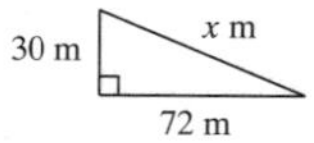

20 What is the value of h? ______________ [1]

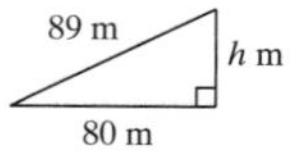

21 Find the value of x. Leave the answer as a square root. ______________ [1]

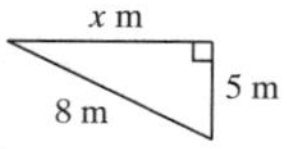

22 Find the value of y. Leave the answer as a square root. ______________ [1]

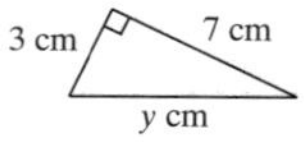

23 What number will complete this Pythagorean triad? 396, 403, ? ______________ [1]

24 What is the value of x? Give the answer to one decimal place. ______________ [1]

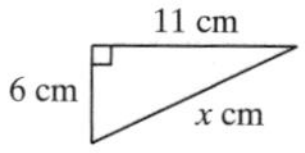

25 What is the value of h? Give the answer to one decimal place. ______________ [1]

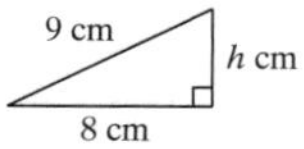

PART C Show all working.

26 a Find the unknown shaded area. [1 mark]

____

b What is the perimeter of the triangle? [2 marks]

____

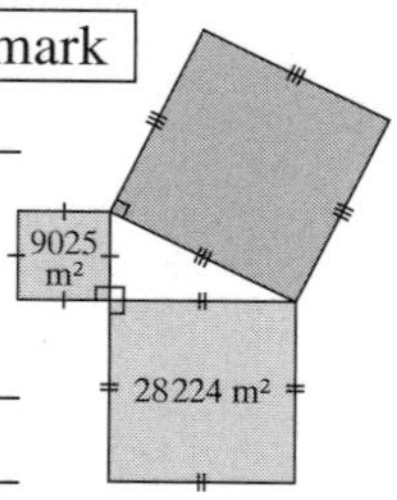

27 What is the length of the diagonal of a rectangle 9 m long and 5 m wide? Give the answer in metres, correct to the nearest centimetre. [2 marks]

____

28 Find the area of this trapezium. [3 marks]

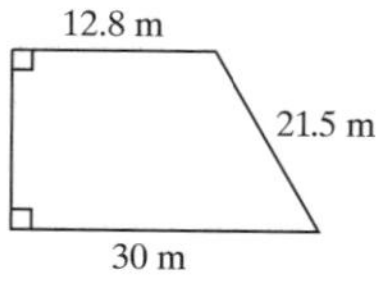

____

29 a Find the value of x. [2 marks]

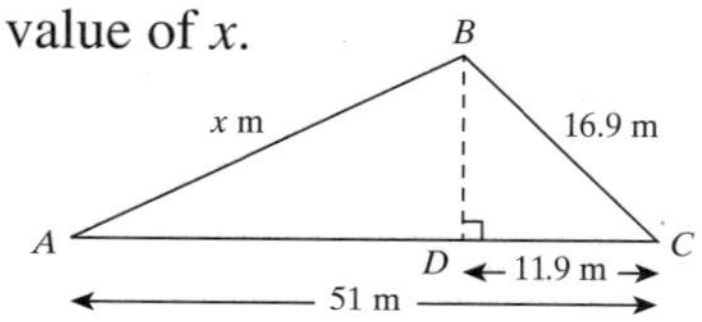

____

b Is ΔABC right-angled? Justify your answer. [2 marks]

____

30 This triangle has area 119.7 m^2.

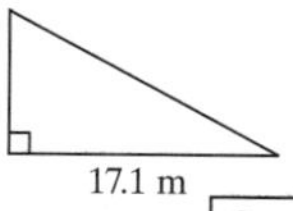

What is the perimeter? [3 marks]

____

Total marks /40

PRETEST

Time allowed: 15 minutes **Total marks 15**

1 What do the angles of a triangle add up to? 1

Ⓐ 180° Ⓑ 200° Ⓒ 360° Ⓓ 380°

2 A triangle with exactly two sides equal is: 1

Ⓐ equilateral Ⓑ isosceles Ⓒ alternate Ⓓ scalene

3 What is the complement of 50°? 1

Ⓐ 30° Ⓑ 40° Ⓒ 50° Ⓓ 130°

4 What type of plane figure has exactly six sides? 1

Ⓐ octagon Ⓑ pentagon Ⓒ heptagon Ⓓ hexagon

5 What do the angles of a quadrilateral add up to? 1

Ⓐ 180° Ⓑ 200° Ⓒ 360° Ⓓ 380°

6 What type of angle measures 125°? 1

Ⓐ acute Ⓑ obtuse Ⓒ reflex Ⓓ straight

7 What type of triangle has all its angles equal? 1

Ⓐ equilateral Ⓑ isosceles Ⓒ right Ⓓ scalene

8 Which single transformation could move ΔABC to $\Delta A'B'C'$? 1

Ⓐ translation Ⓑ rotation Ⓒ reflection Ⓓ enlargement

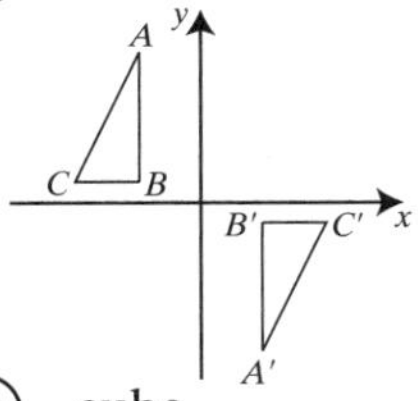

9 Which is **not** a quadrilateral? 1

Ⓐ rhombus Ⓑ trapezium Ⓒ rectangle Ⓓ cube

10 What name could be given to the shaded angle? 1

Ⓐ $\angle SRQ$ Ⓑ $\angle PST$ Ⓒ $\angle RSP$ Ⓓ $\angle TSR$

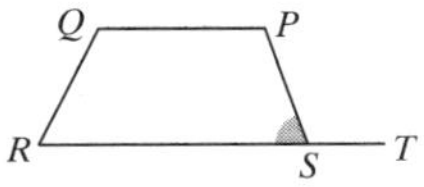

11 Which diagram shows an acute-angled scalene triangle? 1

Ⓐ 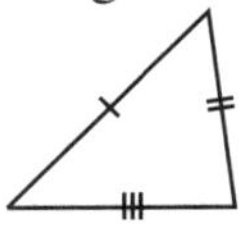 Ⓑ 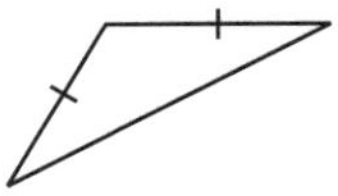 Ⓒ 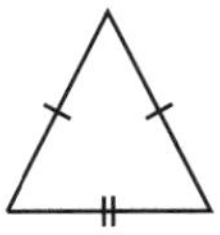 Ⓓ 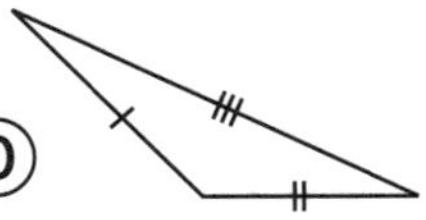

12 What is the value of x? 1

Ⓐ 16 Ⓑ 21 Ⓒ 37 Ⓓ 58

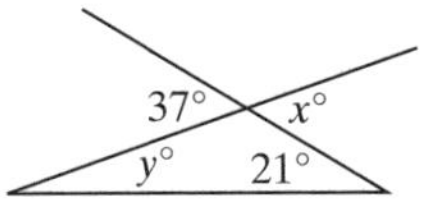

13 What is the value of y? 1

Ⓐ 58 Ⓑ 37 Ⓒ 21 Ⓓ 16

14 If a pair of parallel lines is cut by a transversal, the corresponding angles are: 1

Ⓐ equal Ⓑ complementary Ⓒ supplementary Ⓓ adjacent

15 If a pair of parallel lines is cut by a transversal, the co-interior angles are: 1

Ⓐ equal Ⓑ complementary Ⓒ supplementary Ⓓ adjacent

Total marks /15

LEVEL 1 TEST PART A

Time allowed: 40 minutes

1 Congruent triangles are:

Ⓐ the same shape but not the same size Ⓑ the same size but not the same shape

Ⓒ neither the same shape nor the same size Ⓓ both the same shape and same size [1]

2 Under which transformation might an object and its image **not** be congruent?

Ⓐ reflection Ⓑ rotation Ⓒ translation Ⓓ enlargement [1]

3 What is the symbol for 'is congruent to'?

Ⓐ ≡ Ⓑ ||| Ⓒ ⊥ Ⓓ ∴ [1]

4 Which triangle must be congruent to ΔABC?

Ⓐ 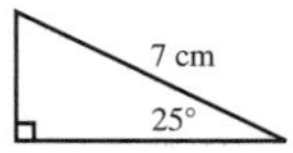

Ⓑ 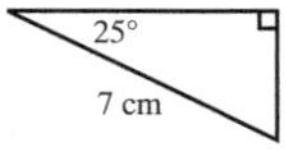

Ⓒ 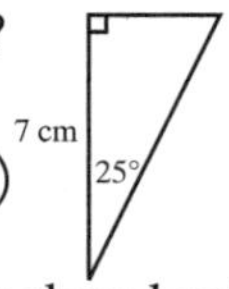

Ⓓ 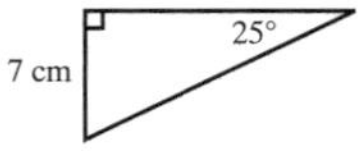

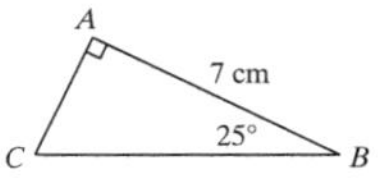

[1]

5 Which quadrilateral is both a rectangle and a rhombus?

Ⓐ kite Ⓑ parallelogram Ⓒ trapezium Ⓓ square [1]

6 Which test proves these triangles congruent?

Ⓐ AAS Ⓑ SAS Ⓒ RHS Ⓓ SSS

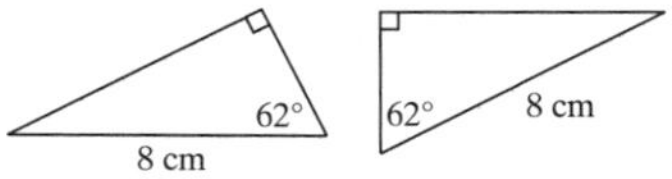

[1]

7 Which of these is **not** a property of a parallelogram?

Ⓐ The diagonals are equal. Ⓑ The diagonals bisect each other.

Ⓒ The opposite sides are parallel. Ⓓ The opposite sides are equal. [1]

8 IF *JKLM* is congruent to *PQRS*, what is the size of $\angle LKJ$?

Ⓐ 110° Ⓑ 105° Ⓒ 85° Ⓓ 60°

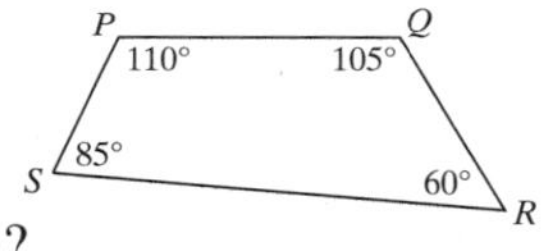

[1]

9 Which of these quadrilaterals always has diagonals that are equal in length?

Ⓐ rhombus Ⓑ rectangle Ⓒ trapezium Ⓓ none of these [1]

10 Which of these quadrilaterals always has diagonals that meet at right angles?

Ⓐ rhombus Ⓑ rectangle Ⓒ trapezium Ⓓ none of these [1]

11 Which quadrilateral could have only one pair of opposite sides parallel?

Ⓐ rhombus Ⓑ rectangle Ⓒ trapezium Ⓓ none of these [1]

12 Which pair of triangles must be congruent?

Ⓐ 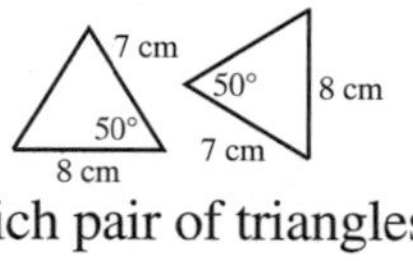

Ⓑ 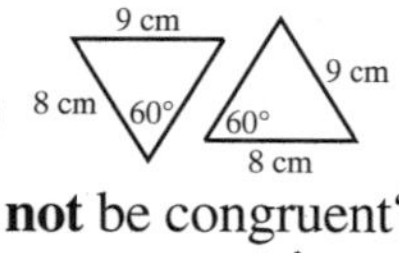

Ⓒ 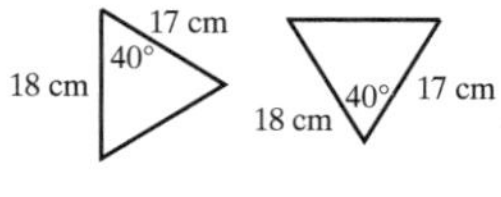

Ⓓ 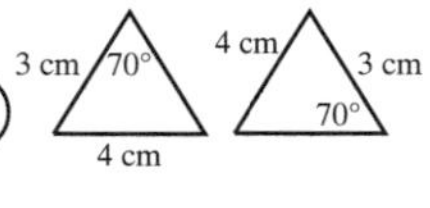

[1]

13 Which pair of triangles might **not** be congruent?

Ⓐ 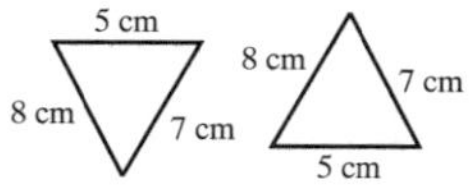

Ⓑ 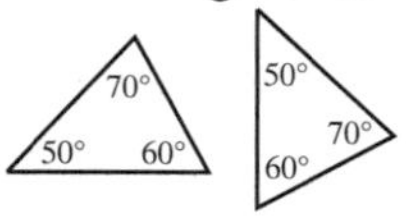

Ⓒ 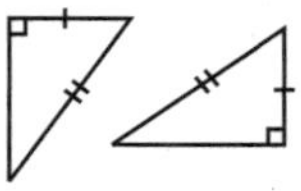
Ⓓ 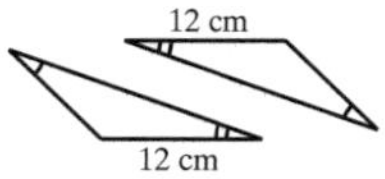

[1]

14 What is the length of the longest side of any triangle congruent to ΔLMN?

Ⓐ 15 m Ⓑ 28 m Ⓒ 17 m Ⓓ not enough information

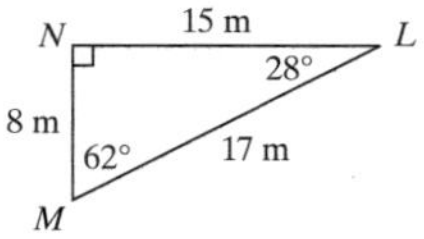

[1]

15 Referring to the triangle in question 14, if ΔPQR is congruent to ΔLMN then $\angle PQR$ =

Ⓐ 90° Ⓑ 62° Ⓒ 28° Ⓓ not enough information [1]

PART B Write answers only in the space provided.

16 Complete: If two shapes are congruent then the matching angles are ________________. 1

17 A triangle with sides 6 cm, 8 cm and 10 cm could be congruent to a triangle with sides of length 3 cm, 4 cm and 5 cm. True or false? ________________ 1

18 If two shapes are congruent they must be triangles. True or false? ________________ 1

19 Which test is used to show these triangles are congruent? ________________ 1

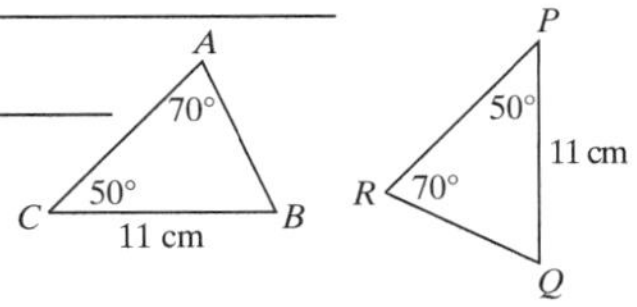

20 What side of ΔPQR corresponds to AC? ________________ 1

21 What is the size of $\angle PQR$? ________________ 1

22 The diagonals of a rhombus bisect the angles through which they pass. True or false? ________________ 1

23 Both pairs of opposite sides of a kite are equal. True or false? ________________ 1

24 IF ΔDFE is congruent to ΔHGI, which side would correspond to HI? ________________ 1

25 Complete: If a rhombus has one angle of 90° it must be a ________________. 1

PART C Show all working.

26 In $PQRS$, PQ is parallel to SR and PS is parallel to QR.

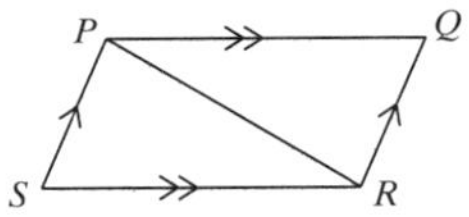

a What type of quadrilateral is $PQRS$? 1 mark

____

b Why does $\angle QPR = \angle SRP$? 1 mark

____

c Why does $\angle PRQ = \angle RPS$? 1 mark

____

d Why does side PR of ΔPQR equal side RP of ΔRSP? 1 mark

____

e Which congruence test shows that triangles PQR and RSP are congruent? 1 mark

____

f Briefly explain why it follows that $\angle PQR = \angle RSP$. 1 mark

____

g What property of this special quadrilateral does this prove? 1 mark

____

____

27 In this quadrilateral $AD = CD$ and $AB = CB$. The diagonals meet at X.

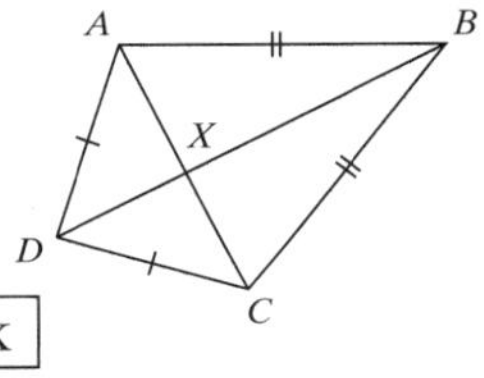

a Which test shows ΔABD is congruent to ΔCBD? 1 mark

____

b Name an angle that must be equal to $\angle ADB$. 1 mark

____

c Which test then shows ΔADX is congruent to ΔCDX? 1 mark

____

d What is the size of $\angle AXD$? Justify your answer. 2 marks

____

____

28 $ABCD$ is a parallelogram. $AB = BC$ and $\angle DBC = 37°$.

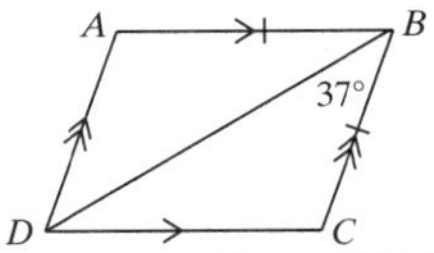

a What special type of parallelogram is $ABCD$? 1 mark

____

b Find the size of $\angle BAD$, giving reasons. 2 marks

____

____

Total marks /40

LEVEL 2 TEST PART A

Time allowed: 40 minutes

1 If two triangles are congruent then the corresponding angles are:
(A) equal (B) complementary (C) supplementary (D) alternate [1]

2 Which angle of ΔPSR corresponds to $\angle PRQ$ of ΔPQR?
(A) $\angle PRS$ (B) $\angle PSR$ (C) $\angle SPR$ (D) $\angle QPS$ [1]

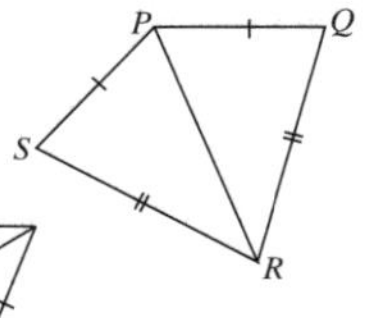

3 What is the value of x?
(A) 30 (B) 60 (C) 90 (D) 120 [1]

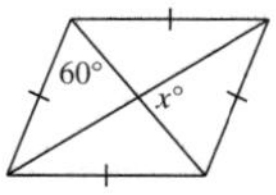

4 Which test would you use to show these two triangles congruent?
(A) AAS (B) RHS (C) SAS (D) SSS [1]

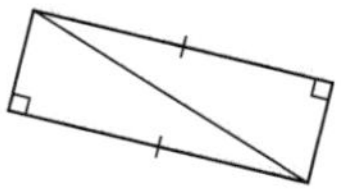

5 Which is **not** necessarily a parallelogram?
(A) square (B) rhombus (C) rectangle (D) trapezium [1]

6 Which pair of triangles must be congruent?
(A) 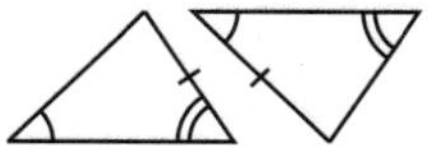 (B) 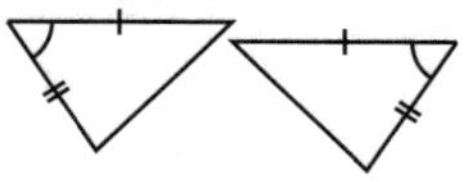 (C)  (D) 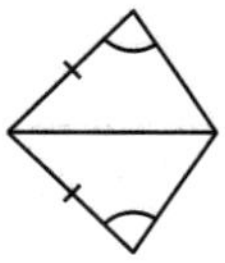 [1]

7 If $\Delta LMN \equiv \Delta YXZ$ then side LN corresponds to:
(A) XY (B) YX (C) XZ (D) YZ [1]

8 This shows part of a tessellation, where all intervals are equal.
Which shape does **not** form part of the tessellation?
(A) equilateral triangle (B) octagon (C) square (D) hexagon [1]

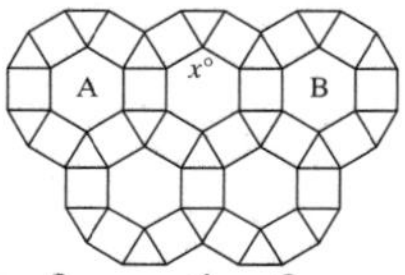

9 In the above tessellation, shape A will **not** map onto shape B under which transformation?
(A) reflection in a horizontal line (B) reflection in a vertical line (C) translation (D) rotation [1]

10 What is the value of x in the above tessellation?
(A) 100 (B) 110 (C) 120 (D) 140 [1]

11 Which is **not** necessarily true for a parallelogram?
(A) The opposite sides are equal. (B) The opposite angles are equal.
(C) The opposite sides are parallel. (D) The diagonals are equal. [1]

12 Which triangle is congruent to ΔABC?

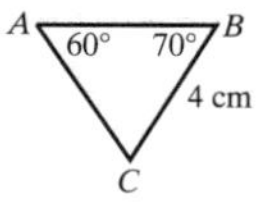

(A) 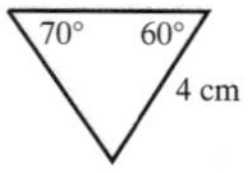

(B) 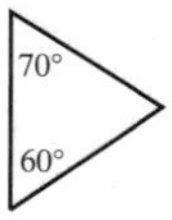

(C) 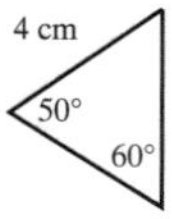

(D) 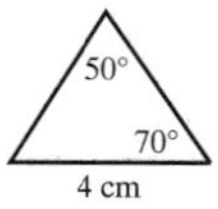

[1]

13 Which must be congruent?
(A) any two isosceles triangles with equal sides equal (B) any two squares
(C) any two right-angled triangles with equal hypotenuses (D) any two circles with the same radii [1]

14 These two quadrilaterals are congruent. So $ABCF \equiv$?
(A) $FCDE$ (B) $CDEF$ (C) $EDCF$ (D) $DEFC$ [1]

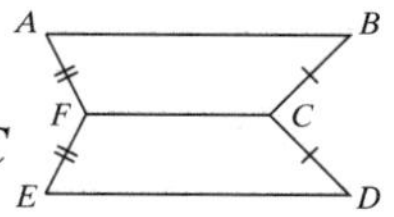

15 $BC = DC$. Which extra information will mean $\Delta ABC \equiv \Delta ADC$?
I: $\angle ABC = \angle ADC$ II: $\angle BCA = \angle DCA$
(A) I only (B) II only (C) both I and II (D) neither I nor II [1]

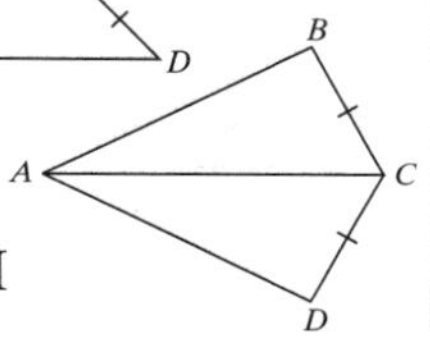

PART B Write answers only in the space provided.

16 Complete: If two triangles are congruent, the corresponding sides are ________________ . 1

17 What is the missing test for congruent triangles? AAS, SAS, SSS, ________________ 1

18 If one angle of a parallelogram measures 70°, what are the sizes of the remaining angles? __________ 1

19 'The diagonals of a parallelogram always meet at right angles.' True or false? ______________ 1

20 Which test shows the two triangles on the right are congruent? ________________ 1

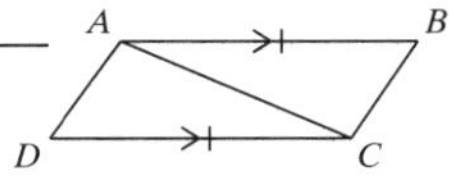

21 Complete for the triangles in question 20. $\Delta ABC \equiv \Delta$________________ 1

22 What type of quadrilateral is *ABCD* in question 20? ________________ 1

23 Referring to the above diagram, if $\angle DAC = 70°$, and $\angle ACD = 30°$, what is the size of $\angle ABC$? ________________ 1

24 If a quadrilateral has all its angles equal, what special quadrilateral must it be?______________ 1

25 If a quadrilateral has all its sides equal, what special quadrilateral must it be?______________ 1

PART C Show all working.

26 In this quadrilateral the diagonals meet at right angles at *P*. The diagonals bisect each other.

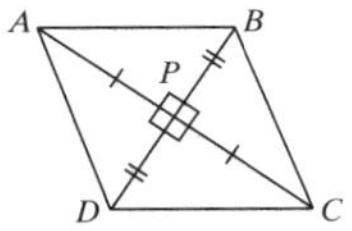

a Which test is used to show $\Delta ABP \equiv \Delta CBP$? 1 mark

____

b Why then, does $AB = CB$? 1 mark

____

c If $\angle BAP = 52°$, what is the size of $\angle PBC$? Give reasons for your answer. 2 marks

____

____

____

____

d Name two other triangles that are congruent to ΔABP. 2 marks

____

____

e What type of quadrilateral is *ABCD*? 1 mark

____

f Briefly explain why the diagonals must also bisect the angles of the quadrilateral. 1 mark

____

____

____

27 Congruent trapeziums have been used to make this pattern.

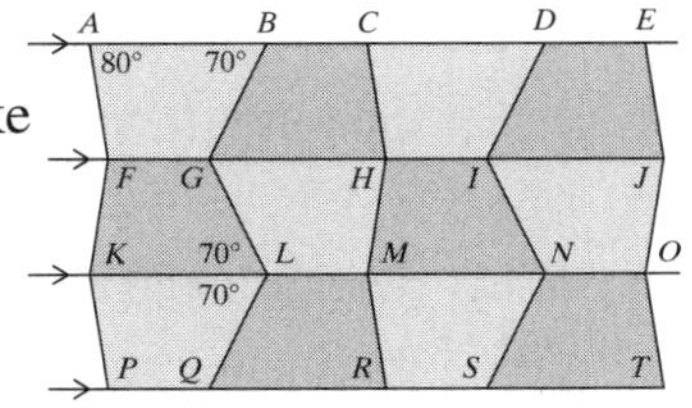

a What side of quadrilateral *HINM* corresponds to *AF* of *ABGF*? 1 mark

____

b What is the size of $\angle JON$? 1 mark

____

c What single transformation will move *GHML* to *QLMR*? 1 mark

____

28 *PQRS* is a square. *PQT* is an equilateral triangle.

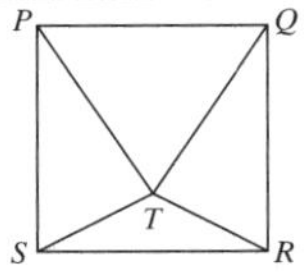

a What is the size of $\angle SPT$? 1 mark

____

b Briefly explain why $PS = PT$. 1 mark

____

____

____

c What is the size of $\angle PST$? 1 mark

____

d What type of triangle is ΔTRS? 1 mark

____

Total marks /40

LEVEL 3 TEST PART A

Time allowed: 40 minutes

1 $\Delta PQR \equiv \Delta SUT$. What is the size of $\angle RPQ$?

(A) 50° (B) 60° (C) 70° (D) not enough information

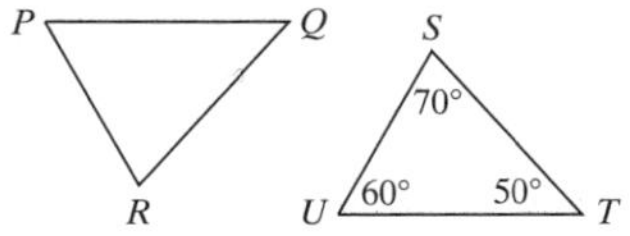

1

2 Which of these triangles is **not** congruent to the other three?

(A) 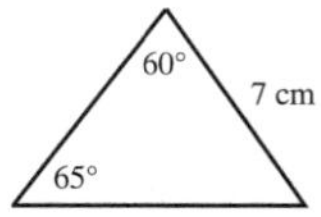

(B) 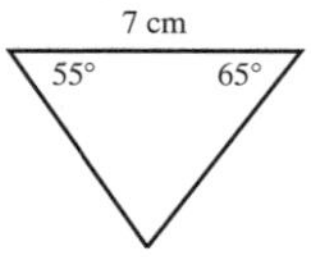

(C) 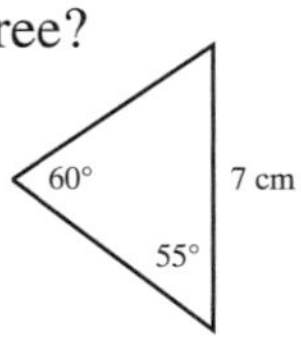

(D) 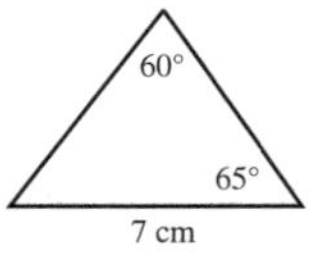

1

3 $x =$

(A) 40 (B) 45 (C) 60 (D) not enough information

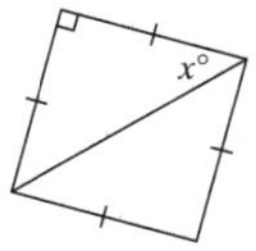

1

4 Two triangles have an angle of the same size and a side of the same length. They **must** be congruent if they also have another equal:

(A) side (B) angle (C) either side or angle (D) neither side nor angle

1

5 Which test shows these two triangles are congruent?

(A) AAS (B) RHS (C) SAS (D) SSS

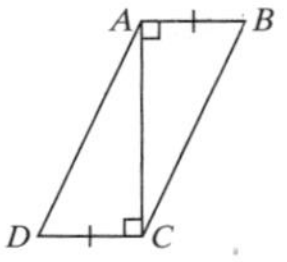

1

6 In the diagram in question 5, which angle corresponds to $\angle DAC$?

(A) $\angle ABC$ (B) $\angle CAB$ (C) $\angle BCA$ (D) $\angle DCB$

1

7 What type of quadrilateral must $ABCD$ in question 5 be?

(A) parallelogram (B) kite (C) rectangle (D) none of these

1

8 Under what transformation will ΔABC map onto ΔADC in the diagram in question 5?

(A) translation (B) reflection in a horizontal line (C) reflection in a vertical line (D) rotation

1

9 Which test shows these two triangles are congruent?

(A) AAS (B) RHS (C) SAS (D) SSS

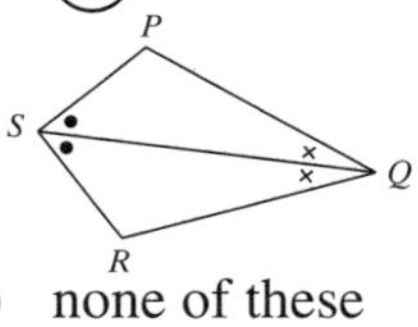

1

10 What type of quadrilateral is $PQRS$ in question 9?

(A) parallelogram (B) kite (C) rectangle (D) none of these

1

11 If, in the quadrilateral in question 9, $\angle PQR = 80°$ and the diagonals meet at T, what is the size of $\angle PTQ$?

(A) 40° (B) 80° (C) 90° (D) 100°

1

12 What is the value of x?

(A) 22 (B) 34 (C) 44 (D) 68

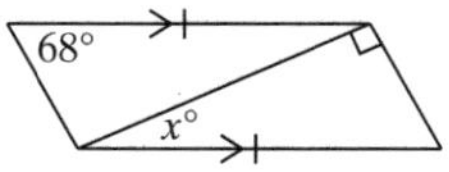

1

13 Which pair of triangles are **not** necessarily congruent?

(A) 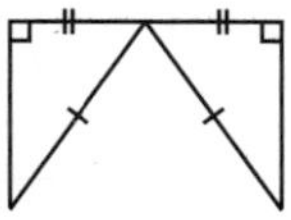

(B) 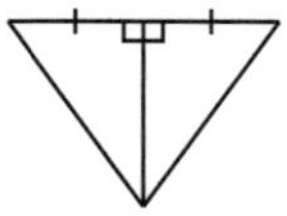

(C) 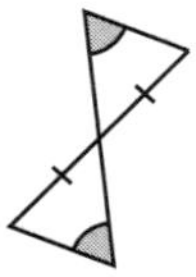

(D) 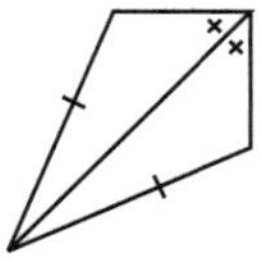

1

14 Which of these triangles must be congruent to ΔPQR?

(A) 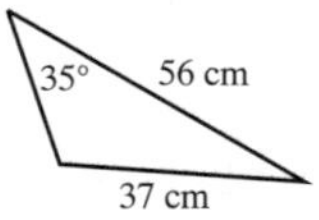

(B) 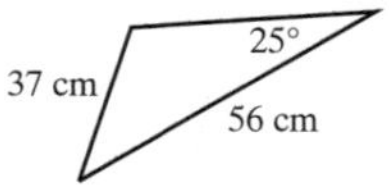

(C) 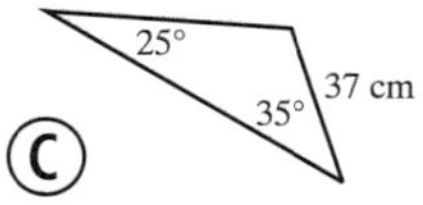

(D) 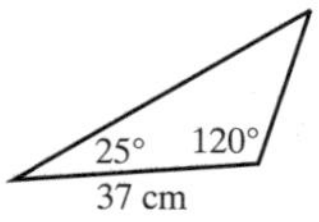

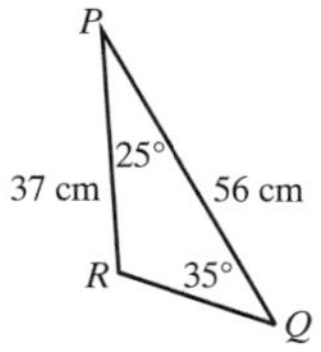

1

15 Which triangles are congruent?

(A) I and II only (B) I and III only

(C) II and III only (D) I, II and III

I 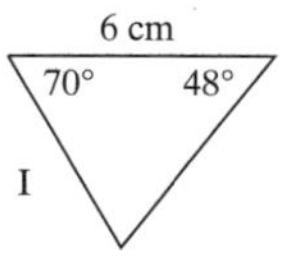

II 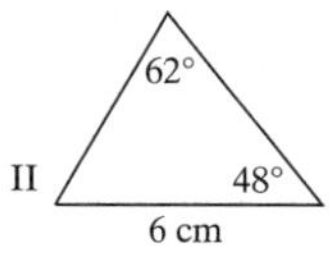

III 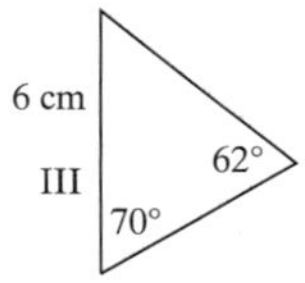

1

PART B Write answers only in the space provided.

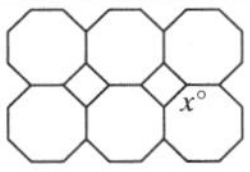

16 This tessellation uses two congruent shapes. What are they?________________ 1

17 What is the value of x in the tessellation above right?________________ 1

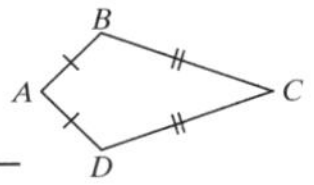

18 $\angle BAD = 120°$ and $\angle ABC = 100°$. What is the size of $\angle BCD$?________________ 1

19 What special name is given to a parallelogram that has a right angle?________________ 1

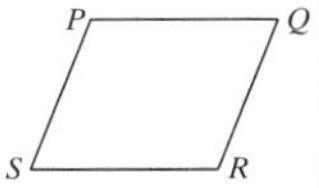

20 $PQRS$ is a rhombus. If $\angle PQR = 76°$, what is the size of $\angle SRP$?________________ 1

21 What is the size of the smallest angle of a quadrilateral if the other three angles are all three times its size?________________ 1

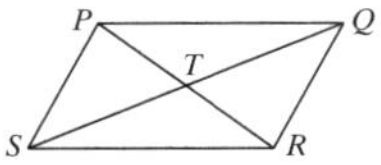

22 $PQRS$ is a parallelogram. T is the point of intersection of the diagonals. Complete $\Delta PQR \equiv \Delta$________________ 1

23 Describe the transformation that maps ΔPQR onto the second triangle in question 22.________________ 1

24 If, in question 22, $PQ = QR$ and $\angle PQS = 32°$, what is the size of $\angle SPQ$?________________ 1

25 If the equal sides of one isosceles triangle are the same length as the equal sides of another isosceles triangle then the triangles must be congruent. True or false?________________ 1

PART C Show all working.

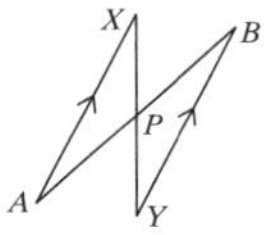

26 AX is parallel to YB.
AB and XY meet at P.
Are the two triangles congruent? Justify your answer. 2 marks

____

____

____

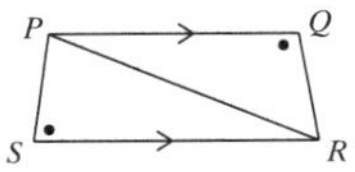

27 Two triangles have a common side as shown. PQ is parallel to SR. $\angle PQR = \angle PSR$

a Name another pair of equal sides or angles. Justify your answer. 2 marks

____

____

b Write the congruence statement for the two triangles and give the test that shows they are congruent. 1 mark

____

c Why must PS be parallel to QR? 2 marks

____

____

____

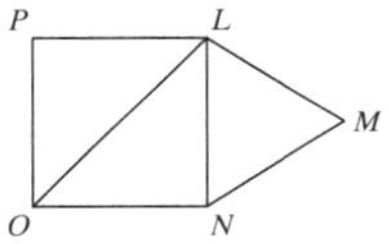

28 $PLNO$ is a square. ΔLMN is equilateral.
What is the size of $\angle OLM$? Justify your answer. 2 marks

____

____

____

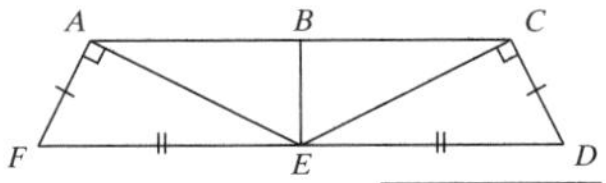

29 a Which test shows ΔAFE is congruent to ΔCDE? 1 mark

____

b What type of triangle is ΔACE? Justify your answer. 2 marks

____

____

c What further single piece of information would mean that $\Delta ABE \equiv \Delta CBE$? Give every possible answer and the test that would be used in each case. 3 marks

____

____

____

Total marks /40

PRETEST

Time allowed: 15 minutes **Total marks 14**

1 How could you describe the chance of getting a 6 when a dice is rolled?

(A) impossible (B) unlikely (C) likely (D) fifty-fifty 1

2 How could you describe the chance of getting a head when a coin is tossed?

(A) impossible (B) unlikely (C) likely (D) fifty-fifty 1

3 If an event is certain to happen its probability will be:

(A) 0 (B) 75% (C) 1 (D) 100 1

4 If an event is impossible its probability will be:

(A) less than 0 (B) 0 (C) 10% (D) 1 1

5 A basket has 24 pegs; six are red, five are white, seven are yellow and the rest are green. Which colour is most likely to occur when a peg is chosen from the basket?

(A) red (B) white (C) yellow (D) green 1

6 This spinner is spun once. On what number is it most likely to stop?

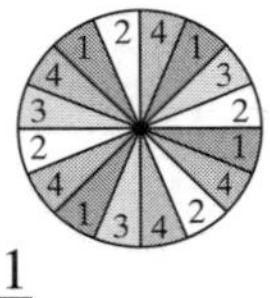

(A) 1 (B) 2 (C) 3 (D) 4 1

7 What is the probability of getting a 5 when a dice is rolled?

(A) $\frac{1}{6}$ (B) $\frac{1}{5}$ (C) $\frac{5}{6}$ (D) $\frac{1}{2}$ 1

8 Which spinner has a fifty-fifty chance of stopping on red?

(A) 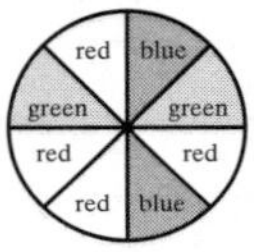

(B) 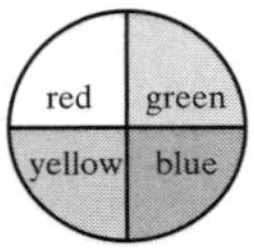

(C) 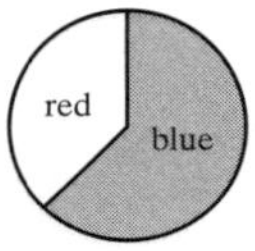

(D) 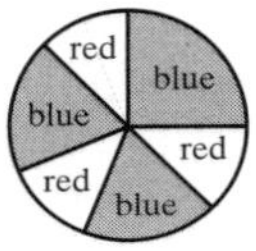

1

9 A box holds five red, three blue and two red marbles. What is the probability of choosing a blue marble?

(A) $\frac{1}{3}$ (B) $\frac{3}{7}$ (C) $\frac{3}{10}$ (D) $\frac{7}{10}$ 1

10 What is the probability of getting a heart when choosing a card from a standard pack?

(A) $\frac{1}{2}$ (B) $\frac{1}{4}$ (C) $\frac{1}{13}$ (D) $\frac{1}{52}$ 1

11 What is the chance of choosing a white ball from a box holding three white, seven red and five black balls?

(A) $\frac{1}{3}$ (B) $\frac{1}{4}$ (C) $\frac{1}{5}$ (D) $\frac{1}{7}$ 1

12 What is the probability of choosing a pink ticket from a hat holding 100 tickets if 40 are green and the rest are pink?

(A) $\frac{1}{60}$ (B) $\frac{1}{40}$ (C) 0.4 (D) 0.6 1

13 What is the probability of choosing a blue balloon from a packet holding four red, six blue, three white, two green and five yellow balloons?

(A) 20% (B) 30% (C) 40% (D) 60% 1

14 What is the probability of rolling a 7 on a standard dice?

(A) 0 (B) $\frac{1}{6}$ (C) $\frac{1}{7}$ (D) $\frac{7}{6}$ 1

Total marks /14

LEVEL 1 TEST PART A

Time allowed: 40 minutes

1 If the probability of choosing a red ball is $\frac{2}{3}$, what is the probability of **not** getting a red ball? 1

(A) $\frac{1}{2}$ (B) $\frac{1}{3}$ (C) $\frac{2}{3}$ (D) not enough information

2 A box holds thirteen white, nine red, eight blue and six yellow golf tees. What is the probability of choosing a red or yellow tee? 1

(A) $\frac{5}{12}$ (B) $\frac{5}{7}$ (C) $\frac{7}{12}$ (D) $\frac{7}{18}$

3 This Venn diagram shows the number of students who caught a bus or train to school. How many students caught a bus? 1

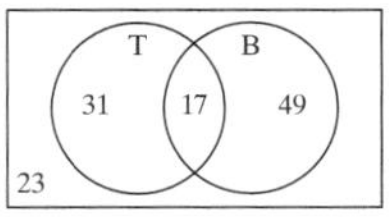

(A) 17 (B) 48 (C) 49 (D) 66

4 Referring to the Venn diagram above, how many students caught a bus or a train but **not** both? 1

(A) 17 (B) 23 (C) 80 (D) 97

5 Referring to the Venn diagram, what is the probability that a student caught both a train and a bus? 1

(A) $\frac{17}{49}$ (B) $\frac{17}{80}$ (C) $\frac{17}{97}$ (D) $\frac{17}{120}$

6 A hat holds 100 tickets, 40 of which are green, 35 white and the rest blue. What is the probability of drawing a blue ticket? 1

(A) $\frac{1}{4}$ (B) $\frac{2}{5}$ (C) $\frac{7}{20}$ (D) $\frac{17}{20}$

7 A dice is rolled. What is the probability of getting a multiple of 2 or 3 or both? 1

(A) $\frac{5}{6}$ (B) $\frac{2}{3}$ (C) $\frac{1}{3}$ (D) $\frac{1}{2}$

8 This spinner is spun once. What is the probability of **not** getting 3? 1

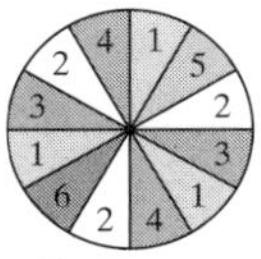

(A) $\frac{5}{6}$ (B) $\frac{4}{5}$ (C) $\frac{2}{3}$ (D) $\frac{3}{4}$

9 What is the probability of getting at least 3 on the spinner in question 8? 1

(A) $\frac{1}{2}$ (B) $\frac{1}{3}$ (C) $\frac{2}{3}$ (D) $\frac{5}{12}$

10 What is the probability of getting at most 3 on the spinner in question 8? 1

(A) $\frac{1}{2}$ (B) $\frac{2}{3}$ (C) $\frac{1}{3}$ (D) $\frac{7}{12}$

11 A card is chosen from a standard pack. What is the probability that it is an ace or a king? 1

(A) $\frac{1}{4}$ (B) $\frac{1}{13}$ (C) $\frac{2}{13}$ (D) $\frac{1}{26}$

12 A spinner can land on 4, 5 or 6. If $P(4) = \frac{1}{4}$ and $P(5) = \frac{1}{5}$ what is $P(6)$? 1

(A) $\frac{1}{6}$ (B) $\frac{11}{20}$ (C) $\frac{7}{9}$ (D) $\frac{9}{20}$

13 There is a 35% chance of hitting a target. What is the chance of missing the target? 1

(A) 35% (B) 45% (C) 55% (D) 65%

14 A basket holds ten red, five blue, seven green and eight yellow pegs. If one peg is taken, what is the probability that it is **not** blue? 1

(A) $\frac{2}{3}$ (B) $\frac{3}{4}$ (C) $\frac{4}{5}$ (D) $\frac{5}{6}$

15 The numbers from 1 to 20 (inclusive) are written on separate cards. If one card is taken at random, what is the probability that it is even and prime? 1

(A) 0 (B) $\frac{1}{18}$ (C) $\frac{1}{20}$ (D) $\frac{2}{5}$

PART B Write answers only in the space provided.

16 The probability of winning a raffle is $\frac{1}{1000}$. What is the probability of **not** winning?_______________ 1

17 What is the probability of **not** getting a 6 when a dice is rolled?_______________ 1

18 There are six red, nine blue and eight green balls in a box. What is the probability of selecting a ball that is either red or green?_______________ 1

19 What is the probability of selecting a diamond and a queen when a card is chosen from a standard pack of playing cards?_______________ 1

20 What is the probability of selecting a diamond or a queen or both when a card is chosen from a standard pack of playing cards?_______________ 1

21 Complete: The sum of all the probabilities of an event must be _______________. 1

22 How many whole numbers are between 1 and 10, both inclusive?_______________ 1

23 A day of the week is chosen at random. What is the probability that it does **not** begin with T?_______________ 1

24 When a dice is rolled, rolling an even number and rolling a number greater than 5 are mutually exclusive events. True or false?_______________ 1

25 What is the probability of rolling at least 4 when a dice is rolled?_______________ 1

PART C Show all working.

26

Corey has these cards; six are black and six are red. If one card is chosen at random, what is the probability that:

a it shows 6? 1 mark

____

b it is both red and shows 6? 1 mark

____

c a card that shows 6 is red? 1 mark

____

d the number on the card is at least 6? 1 mark

____

27 The Venn diagram shows the number of students who play a piano, guitar or violin.

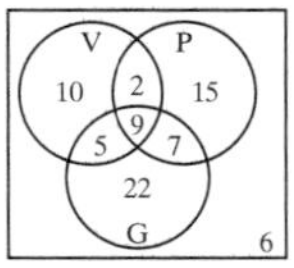

a How many students can play all three instruments? 1 mark

____

b How many students play none of the instruments? 1 mark

____

c How many can play the piano? 1 mark

____

d How many play exactly one instrument? 1 mark

____

e What is the probability that a student who plays piano can also play guitar? 1 mark

____

f What is the probability that a student who plays an instrument, plays piano or violin or both? 1 mark

____

28 One hundred people travelled to work or school by either train or bus.

	Train	Bus
Work	27	16
School	21	

a The number who travelled to school by bus is missing from the table. What should it be? 1 mark

____

b What is the probability that a person travelled by bus? 1 mark

____

c What is the probability that a person who travelled to work caught a bus? 1 mark

____

d What is the probability that a person who travelled by bus was going to work? 1 mark

____

e What is the probability that a person either travelled to school by train or to work by bus? 1 mark

____

Total marks /40

LEVEL 2 TEST PART A

Time allowed: 40 minutes

1 A dice is rolled. What is the probability of rolling a number that is even or odd?

(A) 0 (B) $\frac{1}{6}$ (C) $\frac{1}{2}$ (D) 1 — 1

2 A number is chosen from between 1 and 10, both exclusively. What is the probability that the number is 7?

(A) $\frac{1}{10}$ (B) $\frac{1}{9}$ (C) $\frac{1}{8}$ (D) $\frac{1}{7}$ — 1

3 A ball is chosen from a box holding 15 red, 20 blue, 18 green, 12 yellow and 25 white balls. What is the probability that the ball is **not** red?

(A) $\frac{2}{3}$ (B) $\frac{3}{4}$ (C) $\frac{4}{5}$ (D) $\frac{5}{6}$ — 1

4 A card is chosen from a standard pack of playing cards. What is the probability that it is either a 3 or a heart, but not both?

(A) $\frac{15}{52}$ (B) $\frac{7}{26}$ (C) $\frac{1}{4}$ (D) $\frac{4}{13}$ — 1

5 What is the probability of getting at least 3 when a dice is rolled?

(A) $\frac{1}{6}$ (B) $\frac{1}{2}$ (C) $\frac{2}{3}$ (D) $\frac{4}{5}$ — 1

Students were asked if they had read either of two books: *Blue Depths* and *Wild Safari*. The results are shown in the Venn diagram which is used in questions 6 to 8.

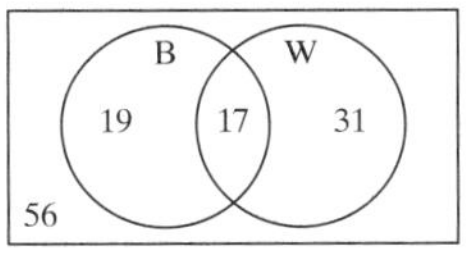

6 How many of the students had read at least one of the books?

(A) 67 (B) 56 (C) 50 (D) 17 — 1

7 What is the probability that a student had **not** read *Blue Depths*?

(A) $\frac{56}{123}$ (B) $\frac{31}{123}$ (C) $\frac{87}{123}$ (D) $\frac{104}{123}$ — 1

8 What is the probability that a student who had read *Wild Safari* had also read *Blue Depths*?

(A) $\frac{17}{48}$ (B) $\frac{17}{36}$ (C) $\frac{17}{31}$ (D) $\frac{17}{19}$ — 1

9 A number is chosen from 1 to 12 (both inclusive). What is the probability that it is **not** prime?

(A) $\frac{7}{12}$ (B) $\frac{2}{3}$ (C) $\frac{7}{10}$ (D) $\frac{1}{2}$ — 1

10 A number from 1 to 100, including 1 and 100, is chosen at random. What is the probability that it is odd, or a multiple of 5, or both?

(A) 50% (B) 55% (C) 60% (D) 70% — 1

11 The probability that a student in a course is male is $\frac{3}{5}$. What is the probability that a student in the course is female?

(A) $\frac{1}{2}$ (B) $\frac{2}{5}$ (C) $\frac{2}{3}$ (D) $\frac{3}{5}$ — 1

12 A box holds pegs that are either red, white or blue. $P(\text{red}) = \frac{1}{3}$ and $P(\text{white}) = \frac{2}{5}$. What is $P(\text{blue})$?

(A) $\frac{1}{3}$ (B) $\frac{2}{5}$ (C) $\frac{2}{15}$ (D) $\frac{4}{15}$ — 1

13 Referring to the table, what fraction of male students were in Year 7?

	Year 7	Year 8
Male	59	71
Female	65	68

(A) $\frac{59}{71}$ (B) $\frac{59}{130}$ (C) $\frac{59}{124}$ (D) $\frac{59}{204}$ — 1

14 Referring to the table above, what is the probability that a female student is in Year 7?

(A) $\frac{65}{124}$ (B) $\frac{65}{133}$ (C) $\frac{124}{133}$ (D) none of these — 1

15 Referring to the table above, what is the probability that a student is **not** in Year 7?

(A) $\frac{124}{139}$ (B) $\frac{124}{263}$ (C) $\frac{263}{139}$ (D) $\frac{139}{263}$ — 1

PART B Write answers only in the space provided.

16 What is the probability of **not** getting a 2 when a dice is rolled?________________ 1

17 What is the probability of getting at most 2 when a dice is rolled?________________ 1

18 What is the probability of getting at least 2 when a dice is rolled?________________ 1

19 There are six green, nine yellow and twelve blue pegs in a box.
What is the probability of **not** choosing yellow?________________ 1

20 'When a card is chosen from a standard pack of playing cards, getting a heart and getting a diamond are mutually exclusive events.' True or false?________________ 1

21 When a card is chosen from a standard pack of playing cards, what is the probability of getting both a spade and a king?________________ 1

22 A spinner has sections in red, blue and green. $P(\text{red}) = \frac{1}{4}$. $P(\text{blue}) = \frac{1}{6}$.
What is $P(\text{green})$?________________ 1

23 The numbers from 1 to 10 are written on 10 cards, one number on each card. What is the probability of choosing a number that is even and greater than 6?________________ 1

24 Twenty students study either chemistry, biology or both. If, altogether, 13 study chemistry and 12 study biology, how many study both?________________ 1

25 A bag holds counters that are black, white and purple. The probability of choosing a black counter is $\frac{1}{4}$ and the probability of **not** getting purple is $\frac{2}{3}$. What is the probability of choosing a white counter?________________ 1

PART C Show all working.

26 Sixty students were asked if they had any of a dog, cat or bird. This Venn diagram shows the number of students who had at least one of those animals.

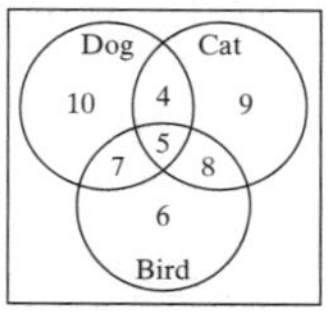

a The number of students who had none of these pets is missing. What should it be? 1 mark

____

b How many had a dog and a cat, but **not** a bird? 1 mark

____

c What is the probability that a student had exactly one of the types of pet? 1 mark

____

d What is the probability that a student had a dog? 1 mark

____

e What is the probability that a student who had a dog also had a bird? 1 mark

____

f What is the probability that a student has neither a dog nor a bird? 1 mark

____

27 This table shows the numbers of junior and senior members of a club.

	Junior	Senior	Total
Male	78	145	
Female	72	93	
Total			

a Complete the totals. 2 marks

b What is the probability that a club member is a junior male? 1 mark

____

c What is the probability that a male club member is a junior? 1 mark

____

d What is the probability that a junior club member is male? 1 mark

____

28 In a class of 23 students, 11 represented the school in athletics, 14 in swimming and six in both.

a Show the information on a Venn diagram. 2 marks

b What is the probability that a student from the class did **not** represent the school in either athletics or swimming? 1 mark

____

c What is the probability that a student represented the school in athletics or swimming, but **not** both? 1 mark

____

Total marks /40

Probability Calculator Allowed

LEVEL 3 TEST PART A

Time allowed: 40 minutes

1 Of 50 students, 17 study art and 19 study music. If eight study both, how many study neither?

(A) 6 (B) 30 (C) 14 (D) 22 [1]

2 Which has the best chance of occurring when a dice is rolled?

(A) an odd number (B) at least 4 (C) at most 3 (D) All have the same chance. [1]

3 What is the probability of choosing neither an ace nor a queen when choosing a card from a standard pack of playing cards?

(A) $\frac{2}{12}$ (B) $\frac{45}{52}$ (C) $\frac{11}{13}$ (D) $\frac{7}{52}$ [1]

4 A dice is rolled. What is the probability that it shows an odd number or a number greater than 5?

(A) $\frac{1}{6}$ (B) $\frac{1}{2}$ (C) $\frac{2}{3}$ (D) $\frac{5}{6}$ [1]

5 The probability of getting a number less than 7 on a spinner is 42%. What is the probability of getting a number more than 7?

(A) 16% (B) 42% (C) 58% (D) not enough information [1]

6 A two-digit number is randomly selected. What is the probability that it is a multiple of both 4 and 7?

(A) $\frac{1}{30}$ (B) $\frac{1}{33}$ (C) $\frac{1}{45}$ (D) $\frac{1}{50}$ [1]

7 If a card is chosen from a standard pack, which are mutually exclusive events?

(A) heart and red (B) club and king (C) diamond and spade (D) queen and black [1]

Some people were asked whether they preferred a cat or dog as a pet. The results are shown in the table, which is used for questions 8, 9 and 10.

	Cat	Dog
Male	28	42
Female	57	53

8 How many people preferred a cat?

(A) 28 (B) 70 (C) 81 (D) 85 [1]

9 What is the probability that a female preferred a dog?

(A) $\frac{53}{95}$ (B) $\frac{53}{110}$ (C) $\frac{53}{180}$ (D) $\frac{57}{85}$ [1]

10 What is the probability that a person who preferred a dog was female?

(A) $\frac{53}{95}$ (B) $\frac{53}{110}$ (C) $\frac{53}{180}$ (D) $\frac{57}{85}$ [1]

11 The probability of winning a particular game is 0.35 and the probability of drawing, tying or not getting a result is 0.35. What is the probability of losing the game?

(A) 0.4 (B) 0.45 (C) 0.3 (D) 0.35 [1]

12 A bag has counters in black, white and gold. $P(\text{white}) = \frac{2}{3}$ and $P(\text{black}) = \frac{1}{4}$. If there are 12 gold counters, how many are black?

(A) 28 (B) 36 (C) 48 (D) not enough information [1]

13 Of 60 students, 39 play tennis and 28 play hockey. If 6 play neither sport, how many play both?

(A) 7 (B) 13 (C) 14 (D) 26 [1]

14 The numbers from 1 to 12 are written on 12 cards (one number on each card). What is the probability of choosing a card that is even or prime, but **not** both?

(A) $\frac{2}{3}$ (B) $\frac{3}{4}$ (C) $\frac{5}{6}$ (D) $\frac{11}{12}$ [1]

15 Some adults and children attended a park. The probability that a person selected at random is an adult male is $\frac{1}{3}$ and the probability that she is an adult female is $\frac{2}{5}$. What is $P(\text{child})$?

(A) $\frac{4}{15}$ (B) $\frac{1}{4}$ (C) $\frac{1}{3}$ (D) $\frac{1}{5}$ [1]

PART B Write answers only in the space provided.

16 What is the probability of **not** getting a 6 when a dice is rolled?________________ [1]

17 What is the probability of getting a jack or a 3 when a card is selected from a standard pack of playing cards? ________________ [1]

18 The numbers from 1 to 25, inclusive, are written on separate cards. What is the probability of selecting a card that shows at least 19?________________ [1]

19 A box holds pegs, half of which are red. One-quarter are blue, one-tenth are green and the rest are yellow. What is the probability of choosing a green peg from the box?________________ [1]

20 What is the probability of selecting a yellow ball from the box in question 19?________________ [1]

21 What event is the complement of rolling a number greater than 1 when a dice is rolled?________________ [1]

22 The numbers 1 to 9 are written on nine separate cards, one number on each card. What is the probability of getting an odd number or a number greater than 6 or both?________________ [1]

23 The numbers 1 to 9 are written on nine separate cards, one number on each card. What is the probability of getting a number that is both odd and greater than 6?________________ [1]

24 The numbers 1 to 9 are written on nine separate cards, one number on each card. What is the probability of getting an odd number or a number greater than 6 but **not** both?________________ [1]

25 The numbers from 10 to 24 (exclusive) are written on separate cards. What is the probability of choosing a card that is at most 20?________________ [1]

PART C Show all working.

26 The Venn diagram shows the numbers of students who have won ribbons in each of three colours: red, yellow and blue.

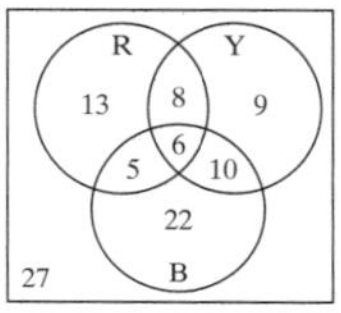

a How many students have both yellow and blue ribbons? [1 mark]

____

b What is the probability that a student from the group has all three colour ribbons? [1 mark]

____

c How many students have ribbons in at least one colour? [1 mark]

____

d How many students have ribbons in at most one colour? [1 mark]

____

e What is the probability that a student has no red ribbons? [1 mark]

____

f What is the probability that a student with a red ribbon also has a blue ribbon? [1 mark]

____

g What is the probability that a student with a ribbon has a red, yellow or both? [1 mark]

____

27 A box holds red, green and blue marbles. The probability of choosing a red marble is $\frac{1}{3}$ and the probability of choosing a green marble is $\frac{2}{5}$.

If there are 24 green marbles, how many blue marbles are there? [2 marks]

____

____

____

28 A survey was taken of 112 people to see if they owned a car and had a driver's licence.

	Owns a car	Does not own a car
Has a licence	44	
Does not have a licence	9	28

The table shows some of the results.

a How many of the people surveyed have a driver's licence? [1 mark]

____

b Complete the Venn diagram. [2 marks]

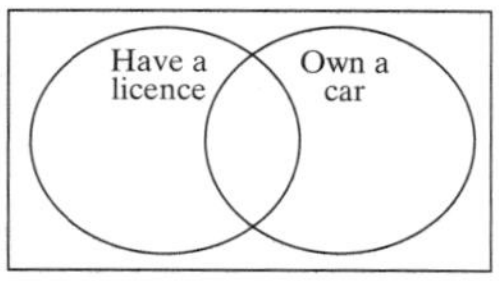

c What is the probability that a person from the group with a licence does **not** own a car? [1 mark]

____

d What is the probability that a person surveyed has a licence or owns a car but **not** both? [1 mark]

____

e Based on these results, about how many of 1000 people would own a car? [1 mark]

____

Total marks /40

PRETEST

Time allowed: 15 minutes **Total marks 15**

1 What is the average of 7, 8 and 21?

(A) 9 (B) 10 (C) 11 (D) 12 [1]

2 What is the range of the scores 6, 3, 2, 7, 2?

(A) 2 (B) 3 (C) 4 (D) 5 [1]

3 What is the mean of the scores 6, 3, 2, 7, 2?

(A) 2 (B) 3 (C) 4 (D) 5 [1]

4 What is the median of the scores 6, 3, 2, 7, 2?

(A) 2 (B) 3 (C) 4 (D) 5 [1]

5 What is the mode of the scores 6, 3, 2, 7, 2?

(A) 2 (B) 3 (C) 4 (D) 5 [1]

6 Which shows the proper tally marks for a score of 13?

(A) ||||| ||||| ||| (B) 𝍸 𝍸 𝍸 | (C) 𝍸 𝍸 ||| (D) 𝍸 ||| [1]

Students voted for their favourite colour. The results are shown in this graph which is used in questions 7 to 9.

Favourite colour

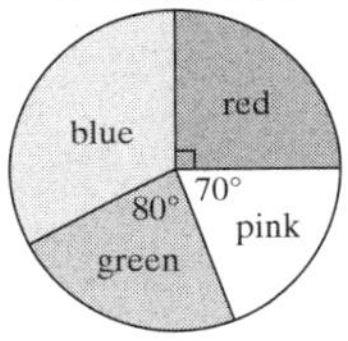

7 What name is given to this type of graph?

(A) line graph (B) sector graph (C) column graph (D) bar graph [1]

8 What size is the angle for blue?

(A) 120° (B) 140° (C) 150° (D) 160° [1]

9 If 72 students voted altogether, how many voted for red?

(A) 16 (B) 18 (C) 20 (D) 24 [1]

This stem-and-leaf plot shows the heights of all animals in a shelter. It is used in questions 10 and 11.

Heights of animals (cm)

14	3	6	8					
15	2	4	4	5	7	9		
16	0	3	4	6	8	8	8	
17	1	1	3	5	5	7	9	
18	0	4	7					
19	1							

10 How many animals are in the shelter?

(A) 26 (B) 27 (C) 32 (D) 33 [1]

11 How tall was the tallest animal?

(A) 814 cm (B) 9 cm (C) 19 cm (D) 191 cm [1]

12 This dot plot shows the marks students received in a quiz. What was the mode?

Marks in the quiz

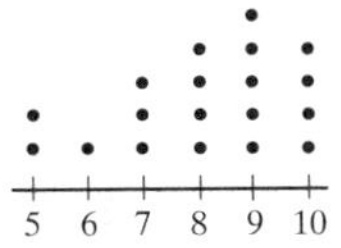

(A) 4 (B) 8 (C) 9 (D) 10 [1]

13 What was the median for the dot plot in question 12?

(A) 7 (B) 7.5 (C) 8 (D) 9 [1]

14 This graph shows the results of an election. Who won?

Election results

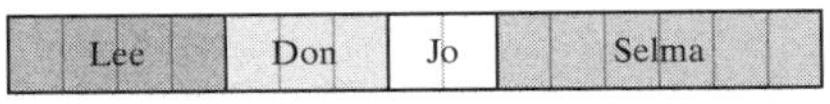

(A) Lee (B) Don (C) Jo (D) Selma [1]

15 Ninety people voted in the above election. Which candidate did exactly 24 people vote for?

(A) Lee (B) Don (C) Jo (D) Selma [1]

Total marks / 15

Statistics Calculator Allowed

LEVEL 1 TEST PART A

Time allowed: 40 minutes

1 Which is categorical data?
(A) type of pet (B) shoe size of adults (C) weight of babies (D) number of children | 1

2 What is the mean of 12, 16, 35, 19, 10, 8 and 19?
(A) 16 (B) 17 (C) 19 (D) 27 | 1

3 What will happen to the mean in question 2 if the score of 35 is left out?
(A) stay the same (B) increase (C) decrease by 2 (D) decrease by 3 | 1

4 What is the mode of 12, 16, 35, 19, 10, 8 and 19?
(A) 7 (B) 16 (C) 17 (D) 19 | 1

5 What will happen to the mode in question 4 if the score of 35 is left out?
(A) stay the same (B) increase (C) decrease by 11 (D) decrease by 16 | 1

6 What is the median of 12, 16, 35, 19, 10, 8 and 19?
(A) 16 (B) 17 (C) 19 (D) 27 | 1

7 What will happen to the median in question 6 if the score of 35 is left out?
(A) stay the same (B) increase (C) decrease by 2 (D) decrease by 3 | 1

8 What is the range of 12, 16, 35, 19, 10, 8 and 19?
(A) 7 (B) 17 (C) 19 (D) 27 | 1

9 What will happen to the range in question 8 if the score of 35 is left out?
(A) stay the same (B) increase (C) decrease by 11 (D) decrease by 16 | 1

10 Which would be a census?
(A) polling the whole population (B) taking names from a hat
(C) asking every fourth person in a queue (D) randomly asking shoppers at a shopping centre | 1

11 What is an example of data that could be collected by observation?
(A) the age of students (B) the number of pets owned
(C) the colours of cars (D) favourite TV shows | 1

12 Which is an example of continuous numerical data?
(A) the number of students in classes (B) the number of cars at an intersection
(C) the heights of students (D) the colour of cars | 1

13 What would be an outlier in this data set? 18, 21, 25, 19, 2, 17, 23, 18, 28, 20, 19, 18, 21
(A) 18 (B) 21 (C) 2 (D) 28 | 1

14 Which would give a random sample of students from a class?
(A) choosing the whole class (B) picking names from a hat
(C) choosing all the boys (D) choosing all those who are taller than 170 cm | 1

15 Which would **not** change in this data set {140, 180, 200, 210, 210, 760} if the outlier is ignored?
(A) mean (B) mode (C) median (D) range | 1

PART B Write answers only in the space provided.

16 What is the mode for the data: 9, 3, 5, 8, 5, 7, 10, 6, 8, 6 10, 5?________________ 1

17 What is the median for the data: 9, 3, 5, 8, 5, 7, 10, 6, 8, 6 10, 5?________________ 1

18 What is the mean for the data: 9, 3, 5, 8, 5, 7, 10, 6, 8, 6 10, 5?________________ 1

19 What is the range for the data: 9, 3, 5, 8, 5, 7, 10, 6, 8, 6 10, 5?________________ 1

20 What is the outlier in the data set: 86, 95, 83, 88, 94, 42, 85, 90, 99, 88?________________ 1

21 The number of students born in different countries is discrete numerical data. True or false?________ 1

22 What is the range of the scores in the stem-and-leaf plot?________________ 1

23 What is the mode of the scores in the stem-and-leaf plot?________________ 1

24 What is the median of the scores in the stem-and-leaf plot?________________ 1

25 What is the mean of the scores in the stem-and-leaf plot?________________ 1

Scores in examination

Stem	Leaf
8	4 6 7
9	2 3 5 7 8 9
10	1 2 6 6 8
11	3 7 9
12	0

PART C Show all working.

26 A census was taken of the students at a school. They were asked how many books they had read in the past week. The results are shown in the table.

Number of books	Number of students
0	312
1	241
2	95
3	27
4	13
5	1

a Briefly explain what is meant by the fact that this was a census. 1 mark

____

b What type of data is this? 1 mark

____

c How many students are at the school? 1 mark

____

d How many books were read altogether? 1 mark

____

e What is the mode? 1 mark

____

f What is the mean number of books? Give the answer to one decimal place. 1 mark

____

27 These are the weights, in grams, of some babies born in a hospital one week.
3493, 3642, 3075, 3987, 2904, 3214, 3718, 2096, 3357, 2994, 3629, 3167

a What type of data is this? 1 mark

____

b What is the weight of the smallest baby? 1 mark

____

c What is the range? 1 mark

____

d What is the mean? 1 mark

____

e What is the median? 1 mark

____

f If the weight of the smallest baby is left out, find the new range, mean and median and briefly explain any changes. 3 marks

____

____

____

____

____

g Briefly explain why it might or might not be a good idea to leave out the weight of the smallest baby. 1 mark

____

____

____

____

Total marks /40

LEVEL 2 TEST — PART A

Time allowed: 40 minutes

1 Most commonly, the 'average' refers to the:

(A) mean (B) median (C) mode (D) range — 1

2 Which could be found for categorical data?

(A) mean (B) median (C) mode (D) range — 1

Students in a class were asked to spell 10 words. The dot plot shows the results. It is used in questions 3 to 8.

Words correctly spelled

3 What is the range?

(A) 5 (B) 7 (C) 7.5 (D) 8 — 1

4 What is the mode?

(A) 5 (B) 7 (C) 7.5 (D) 8 — 1

5 What is the median?

(A) 5 (B) 7 (C) 7.5 (D) 8 — 1

6 What is the mean?

(A) 5 (B) 7 (C) 7.5 (D) 8 — 1

7 What type of data is this?

(A) discrete numerical (B) continuous numerical (C) categorical (D) none of these — 1

8 What percentage of the students made fewer than three mistakes?

(A) $33\frac{1}{3}\%$ (B) 50% (C) 18% (D) $66\frac{2}{3}\%$ — 1

Consider these scores for questions 9, 10 and 11: 37, 49, 32, 46, 41, 38, 13, 42, 37, 44.

9 What is the outlier?

(A) 37 (B) 13 (C) 49 (D) 44 — 1

10 Which of these will change the most if the outlier is ignored?

(A) mean (B) median (C) mode (D) range — 1

11 Which of these will **not** change if the outlier is ignored?

(A) mean (B) median (C) mode (D) range — 1

This stem-and-leaf plot shows the weights, in kilograms, of students. It is used in questions 12 to 15.

Weights of students (kg)

4	7
5	0 4 6 9
6	1 3 5 5 5 7 7 8
7	0 1 4 6 6 8
8	2 4 9
9	3

12 What is the range?

(A) 46 (B) 56 (C) 65 (D) 67 — 1

13 What is the mode?

(A) 46 (B) 56 (C) 65 (D) 67 — 1

14 What is the median?

(A) 46 (B) 56 (C) 65 (D) 67 — 1

15 Which is closest to the percentage of students that weigh more than 70 kg?

(A) 37% (B) 39% (C) 41% (D) 43% — 1

PART B Write answers only in the space provided.

16 What is the mean of 8, 17 and 23?________________ 1

17 What is the difference between the median and mode of 2, 2, 2, 3, 3, 3, 4, 4, 5, 5, 5, 5, 5?__________ 1

18 If the mean of six scores is 5.5, what is the sum of the scores?________________ 1

19 'Categorical data cannot be ordered.' True or false?________________ 1

20 'Sometimes data can be collected by observation.' True or false?________________ 1

21 What is the lowest score of a data set if the highest score is 62 and the range is 37?__________ 1

The colour of all the cars in a car park was noted and the results are shown in the table which is used for questions 22 to 25.

Colour	Number of cars
White	23
Red	11
Blue	10
Black	13
Green	6
Silver	14

22 What type of data is this?________________________________ 1

23 How many cars were counted?________________ 1

24 What is the most popular colour?________________ 1

25 What fraction of the cars were red?________________ 1

PART C Show all working.

26 **a** Briefly explain the difference between a census and a sample. [1 mark]

____

____

b Give a reason why it might be preferable to survey a sample of the population rather than conduct a census. [1 mark]

____

____

27 This graph shows the time taken by a person to drive to work each morning for a month.

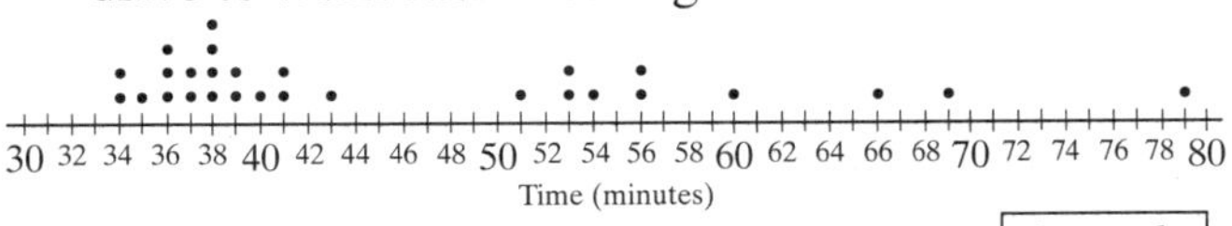

a What is the most common time? [1 mark]

____

b Discuss the spread of the times, pointing out any gaps or clusters. [2 marks]

____

____

c Find the median. [1 mark]

____

d Find the mean of all the times. [1 mark]

____

e Why might it be more appropriate to use the median rather than the mean when analysing the data? [1 mark]

____

____

f Should the time of 79 minutes be ignored? Discuss. [2 marks]

____

____

28 A new garden is being made at a park. A sample of men working at the site is chosen and the length of their normal stride measured. The results, in centimetres, are:

85, 81, 79, 78, 73, 80, 78, 85, 86, 85, 76, 83

a Find the mean length. [1 mark]

____

b Find the modal length. [1 mark]

____

c Find the range. [1 mark]

____

d The results are going to be used to determine the distance between stepping stones that will be placed in the garden. Explain why this might **not** be a good idea. [2 marks]

____

____

____

____

Total marks /40

LEVEL 3 TEST PART A

Time allowed: 40 minutes

1 For the scores 19, 14, 26, 15, 23, 11, 15, 20 and 12, which is **not** equal to 15?

(A) mean (B) median (C) mode (D) range 1

2 For the scores 16, 19, 9, 1, 17, 12 and 17, which of these will decrease if the score of 1 is ignored?

(A) mean (B) median (C) mode (D) range 1

3 For the scores 16, 19, 9, 1, 17, 12 and 17, which will increase the most if the score of 1 is ignored?

(A) mean (B) median (C) mode (D) range 1

4 For the scores 16, 19, 9, 1, 17, 12 and 17, which will stay the same if the score of 1 is ignored?

(A) mean (B) median (C) mode (D) range 1

5 Which is an example of continuous numerical data?

(A) clothing sizes (B) oven temperatures

(C) number of words in sentences (D) types of birds 1

6 Which of these sets of scores has a mean of 45 and a median of 45.5?

(A) 42, 43, 47, 50 (B) 41, 44, 46, 49 (C) 43, 44, 47, 48 (D) 41, 45, 46, 48 1

7 Some students did not do well in a test and had to sit for it again. The second time, each student achieved exactly seven marks more than they did before. Which will **not** have changed?

(A) mean (B) median (C) mode (D) range 1

8 People were asked to rate a documentary from 1 to 10 with 1 being not at all interesting and 10 being extremely interesting. What type of data is that?

(A) continuous numerical (B) discrete numerical (C) categorical (D) cardinal 1

This dot plot shows the number of marks lost by students in a class test. It is used in questions 9 to 14.

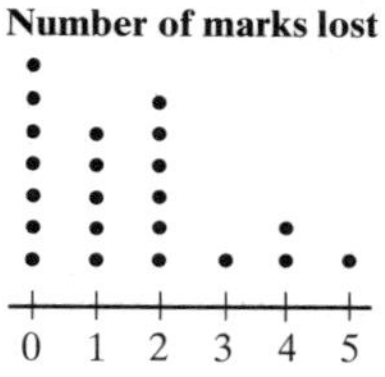

9 How many marks were lost altogether?

(A) 15 (B) 22 (C) 33 (D) 40 1

10 What is the range?

(A) 4 (B) 5 (C) 6 (D) 7 1

11 What is the mode?

(A) 0 (B) 2 (C) 5 (D) 7 1

12 What is the median?

(A) 1 (B) 1.5 (C) 2 (D) 2.5 1

13 What is the mean?

(A) 1 (B) 1.5 (C) 2 (D) 2.5 1

14 What type of data is this?

(A) categorical (B) discrete numerical (C) continuous numerical (D) cardinal 1

15 Which sort of data would be undesirable?

(A) ethical (B) reliable (C) representative (D) biased 1

Statistics Calculator Allowed (Level 3 Test)

PART B Write answers only in the space provided.

16 'Any data that contains numbers must be numerical data.' True or false?________________ 1

17 What is the highest score of a data set if the range is 47 and the lowest score is 78? ______________ 1

18 The mean of eight scores is 17. If a new score of 35 is added, what is the new mean?______________ 1

19 Referring to question 18 above, is the following statement true or false? 'Because the new score of 35 is so far above the mean, it must be an outlier.'________________ 1

20 What is the difference between the mean and mode of 26, 37, 31, 28, 35, 31, 32 and 34?___________ 1

21 'A variable in statistics is something that can be measured or observed to determine change.' True or false?________________ 1

22 'Random means by some particular method.' True or false?_______________ 1

23 What is the mean of $\frac{3}{4}, \frac{5}{8}, \frac{2}{3}$ and $\frac{4}{5}$? _______________ 1

24 What is the median of $\frac{3}{4}, \frac{5}{8}, \frac{2}{3}$ and $\frac{4}{5}$? _______________ 1

25 What is the range of $\frac{3}{4}, \frac{5}{8}, \frac{2}{3}$ and $\frac{4}{5}$? _______________ 1

PART C Show all working.

26 The remaining scores in a data set after an outlier has been discarded are 7, 8, 8, 8 and 12. As a result of ignoring the outlier the mean was reduced by 1.9. What was the outlier? 2 marks

____

____

____

27 An advertisement states that '9 out of 10 people surveyed stated that they would buy this product'. Briefly explain how this statement could be misleading. 2 marks

____

____

28 The ages of students in a course are: 37, 25, 28, 30, 21, 23, 20, 25, 39, 24, 28, 27, 36, 25, 37, 28, 25, 30, 27, 24, 31 and 26.

a What is the range of the ages? 1 mark

____

b Show the ages on a dot plot. 2 marks

c What is the median age? 1 mark

____

d What is the mean age? 1 mark

____

e Comment on the data in regard to any gaps, clusters or outliers. 2 marks

____

____

29 Two different samples of the heights of plants were taken from the same population. One sample of 20 plants had a mean height of 14.7 cm. The second group had 16 plants and the mean height was 23.1 cm.

a Aaron said that the mean height of these plants is 18.9 cm. Explain why Aaron is wrong and find the correct mean. 2 marks

____

____

b Aaron also felt that the differences between the means of the two different samples was too large. Do you think that is correct? What could Aaron do? 2 marks

____

____

____

Total marks /40

Exam Paper 1: Level 1 Non-calculator

Instructions for all parts • Attempt all questions.
Time allowed: 1 hour 40 minutes • Calculators are not allowed.

Part A: Allow about 25 minutes for this part.
Part B: Allow about 30 minutes for this part.
Part C: Allow about 45 minutes for this part. Show all working for this part. **Total marks: 100**

EXAM PAPER 1: LEVEL 1 PART A

Fill in only one circle for each question.

Marks

1 Which unit would be best to measure the volume of a truckload? [1]

(A) km^3 (B) m^3 (C) mm^3 (D) cm^3

2 $4x \times 5x =$ [1]

(A) $9x$ (B) $9x^2$ (C) $20x$ (D) $20x^2$

3 The recurring decimal 3.218 181 818 18… could be written as: [1]

(A) $3.21\dot{8}$ (B) $3.2\dot{1}\dot{8}$ (C) $3.\dot{2}1\dot{8}$ (D) none of these

4 A box holds red, blue and green marbles. One marble is chosen from the box without looking. The probability that the marble is red is $\frac{2}{5}$ and the probability that it is blue is $\frac{2}{5}$. What is the probability that it is green? [1]

(A) $\frac{1}{5}$ (B) $\frac{2}{5}$ (C) $\frac{3}{5}$ (D) $\frac{4}{5}$

5 If $p = 5$, what is the value of $2 + 3p$? [1]

(A) 17 (B) 25 (C) 37 (D) 55

6 How long is it from a quarter to three until 5:55 on the same afternoon? [1]

(A) 2 h 10 min (B) 2 h 40 min (C) 3 h 10 min (D) 3 h 40 min

7 Which of these is irrational? [1]

(A) 450% (B) $\frac{7}{5}$ (C) $0.\dot{2}$ (D) $\sqrt{2}$

8 $3^3 \times 3^4 =$ [1]

(A) 3^{12} (B) 3^7 (C) 9^{12} (D) 9^7

9 What is the area of this parallelogram? [1]

(A) 34 cm^2 (B) 35 cm^2 (C) 50 cm^2 (D) 70 cm^2

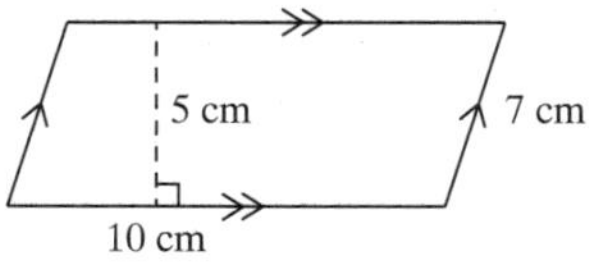

10 What is the result when \$800 is increased by 20%? [1]

(A) \$160 (B) \$820 (C) \$960 (D) \$1000

11 $-10 - 7 \times -2 =$ [1]

(A) -24 (B) -34 (C) 34 (D) 4

12 Surveying which of these groups would be a census of the students in the class? [1]

(A) all the boys in the class
(B) all those who wanted to take part
(C) every second student
(D) every student in the class

Marks

13 $6(2x + 5) =$

(A) $8x + 5$ (B) $8x + 11$ (C) $12x + 5$ (D) $12x + 30$

1

14 What is the volume of this prism?

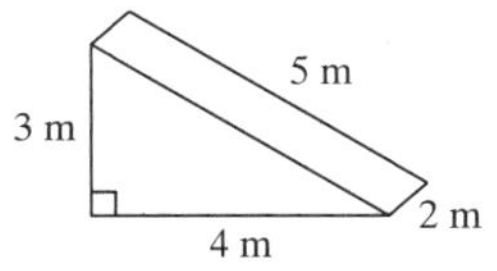

(A) 12 m^3 (B) 14 m^3 (C) 20 m^3 (D) 24 m^3

1

15 What is the ratio of males to females if there are 25 females among 40 people?

(A) 3 : 5 (B) 5 : 3 (C) 5 : 8 (D) 8 : 5

1

16 $3a + 5b - 2a + 4b =$

(A) $a + b$ (B) $a + 9b$ (C) $5a + b$ (D) $5a + 9b$

1

17 The Venn diagram shows the numbers of students taking part in a school play who sing and/or dance. What is the probability that a student in the play both sings and dances?

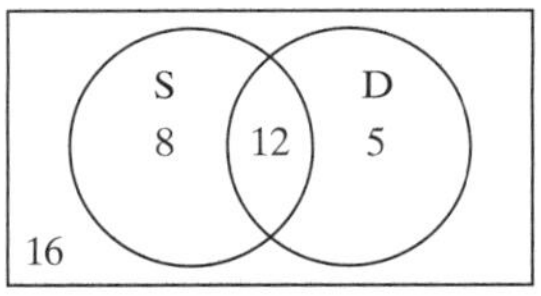

(A) $\frac{8}{41}$ (B) $\frac{12}{41}$ (C) $\frac{13}{41}$ (D) $\frac{25}{41}$

1

18 What is the circumference of a circle of diameter 8 cm?

(A) 4 cm (B) 4π cm (C) 8π cm (D) 16π cm

1

19 How long would it take to travel 90 km at an average speed of 60 km/h?

(A) 40 min (B) 1 h 20 min (C) $1\frac{1}{2}$ h (D) 1 h 40 min

1

20 What is the solution to the equation $5x = 16 - 3x$?

(A) $x = 2$ (B) $x = 4$ (C) $x = 6$ (D) $x = 8$

1

21 Which pair of triangles must be congruent?

(A) 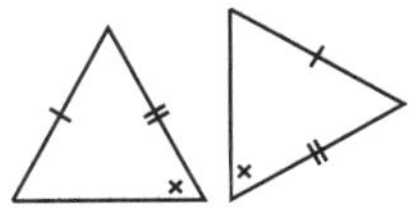 (B) 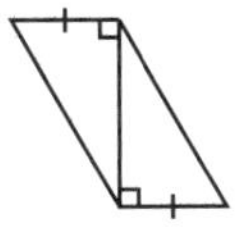 (C) 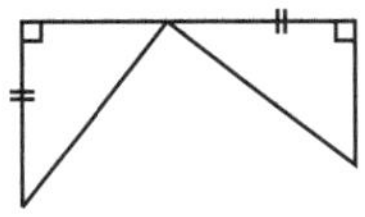 (D) 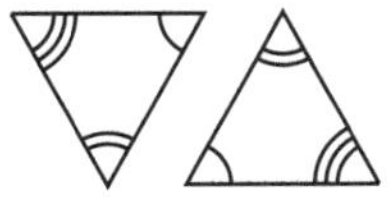

1

22 What is the probability of getting a number that is both odd and greater than 3 when a dice is rolled?

(A) $\frac{1}{6}$ (B) $\frac{1}{3}$ (C) $\frac{1}{2}$ (D) $\frac{5}{6}$

1

23 These are the marks in a test: 1, 16, 17, 17, 18, 18, 18, 18, 19, 20, 20.
Which of these will **not** change when the mark of 1 is left out?

(A) mean (B) median (C) mode (D) range

1

24 How many square centimetres are in 9 m^2?

(A) 300 (B) 900 (C) 30 000 (D) 90 000

1

25 What is the value of x?

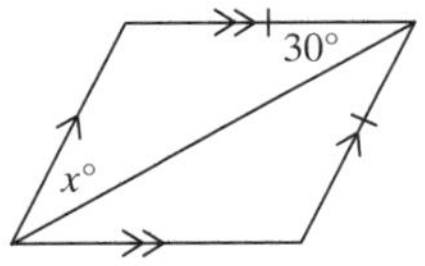

(A) 30 (B) 40 (C) 45 (D) 60

1

Total marks /25

EXAM PAPER 1 : LEVEL 1 PART B

Marks

1 $9^0 =$ ______ 1

2 $-15 \div 3 =$ ______ 1

3 Express $\frac{2}{3}$ as a decimal. ______ 1

4 Factorise $2x - 8$. ______ 1

5 What is the profit if an item bought for \$1200 is sold for \$2100? ______ 1

6 What is the diameter of a circle of radius 7 cm? ______ 1

7 $20x \div 2x =$ ______ 1

8 What is the perimeter of a rhombus if one side is 8 cm long? ______ 1

9 Simplify the ratio $12:6:4$. ______ 1

10 $6a^2 + 4a^2 =$ ______ 1

11 What rate is 210 km in 3 hours? ______ 1

12 $5 - -3 =$ ______ 1

13 What is the solution to the equation $2x - 5 = 7$? ______ 1

14 Expand $2x(x + 5)$. ______ 1

15 $4 - 2 \times 7 =$ ______ 1

16 The line $y = 2x + 3$ will pass through the point (2, 5). True or false? ______ 1

17 If the probability that an arrow hits a target is $\frac{5}{8}$, what is the probability that it misses? ______ 1

18 $5^8 \div 5^3$ in simplest index form is: ______ 1

19 What is the percentage decrease if \$1200 decreases to \$1080? ______ 1

Marks

20 Name a quadrilateral for which the diagonals must be the same length. ____________ [1]

21 What is the time $2\frac{1}{4}$ hours after 11:45 am? ____________ [1]

22 Write $0.\dot{2}\dot{6}$ correct to three decimal places. ____________ [1]

23 $-3(2a - 5b) =$ ____________ [1]

24 If $m = -2$, what is the value of $2m^2$? ____________ [1]

25 If the ratio of cars to trucks to buses is 8 : 3 : 1 and there are 48 cars, how many buses are there? ____________ [1]

26 Which test would be used to show these triangles are congruent? ____________ [1]

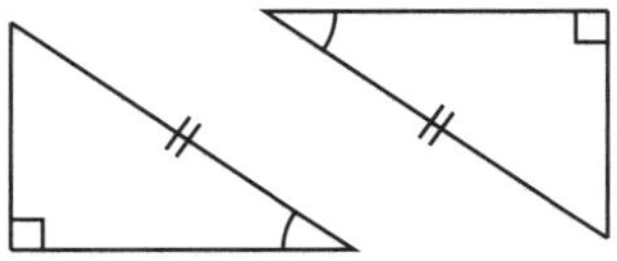

27 $-\frac{2}{3} \times -\frac{4}{5} =$ ____________ [1]

28 What is the area of a kite if one diagonal is 6 cm long and the other 11 cm long? ____________ [1]

29 Write a rule for the values in this table. ____________ [1]

x	0	1	2	3
y	7	8	9	10

30 The area of the triangular face of a triangular prism is 20 cm². If the perpendicular height of the prism is 4 cm, what is the volume? ____________ [1]

Total marks /30

1 Find the value of:

a $-3 \times 5 - 7 \times -2$ [2 marks]

b $100 \div -5 \times 2$ [2 marks]

2 A plane leaves Sydney at 1435 and flies 500 km in a south-westerly direction, landing at 1705.

a At what 12-hour time does the plane leave Sydney? [1 mark]

b How long was the plane in the air? [1 mark]

c What was the average speed of the flight in kilometres per hour? [1 mark]

3 Solve $\frac{x}{3} - 7 = 13$. [2 marks]

4 The scale on a diagram is 1 : 1000.

a Each centimetre on the diagram would represent what actual distance in metres? [1 mark]

b On another diagram 5 cm represents 2 m. What is the scale for that diagram? Give the answer as a ratio in simplest form. [2 marks]

5 Expand and simplify:

a $4(2x - 3y) + 3(x - y)$ [2 marks]

b $n(n + 6) - 2(n + 5)$ [2 marks]

6 A boat bought for $12 000 was later sold for $9000.

a What was the loss? [1 mark]

b What was that loss as a percentage of the cost price? [1 mark]

c The person who bought the boat for $9000 later sold it, making a profit of $33\frac{1}{3}$%. What did that person sell the boat for? [2 marks]

7 **a** A line has been drawn on the number plane below, but not named. What is the equation of the line? [1 mark]

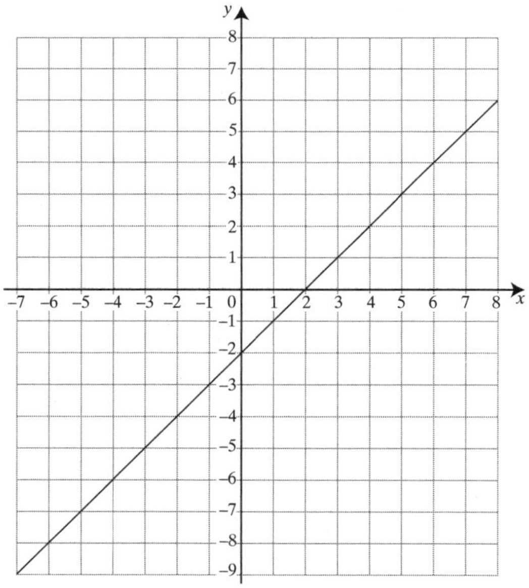

b Complete the table of values for $y = 2x + 2$. [2 marks]

x	0	1	2	3
y				

c Plot the points from the table of values on the number plane, draw and label the line $y = 2x + 2$. [2 marks]

d What is the point of intersection of the two lines? [1 mark]

8 A group of students did a quiz and the marks were: 5, 5, 6, 7, 7, 8, 8, 8, 8, 8.

a What is the range? [1 mark]

b What is the mode? [1 mark]

c What is the median? [1 mark]

d What is the mean? [1 mark]

e What is the probability that a student, chosen at random from the group, scored at least 7? [1 mark]

f Tyson wrote down the scores, but left out one of the 7s. Briefly explain what difference, if any, that will make to the mean, mode, median and range. [2 marks]

9 Write $\left(\frac{3^6 \times 3^2}{3^3 \times 3}\right)^3$ in simplest index form. [3 marks]

10 a Express $\frac{5}{8}$ as a terminating decimal. [1 mark]

b Express $\frac{4}{9}$ as a recurring decimal. [1 mark]

c Are the values in parts **a** and **b** rational or irrational? Justify your answer. [2 marks]

11 a What is the perimeter of this trapezium? [1 mark]

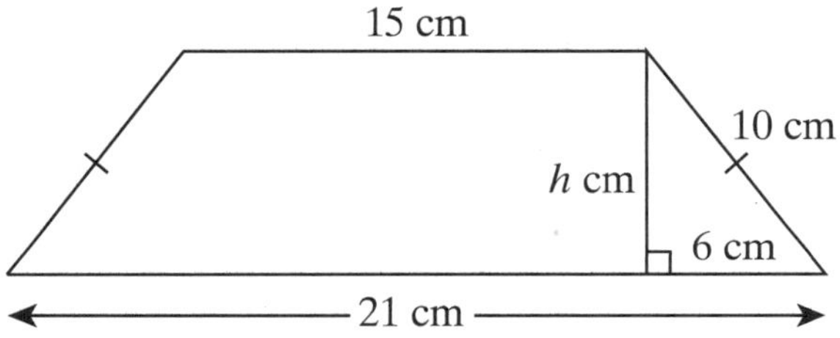

b Find the value of h. [2 marks]

[2 marks]

c Find the area of the trapezium.

Total marks for PART C /45 **Total marks** /100

Exam Paper 2: Level 2 Non-calculator

Instructions for all parts • Attempt all questions.
Time allowed: 1 hour 40 minutes • Calculators are not allowed.

Part A: Allow about 25 minutes for this part.
Part B: Allow about 30 minutes for this part.
Part C: Allow about 45 minutes for this part. Show all working for this part. **Total marks: 100**

EXAM PAPER 2 : LEVEL 2 PART A

Fill in only one circle for each question.

Marks

1 $-3 + 2 \times -6 =$

(A) -15 (B) -9 (C) 6 (D) 30 [1]

2 $5 + 5^0 =$

(A) 5 (B) 6 (C) 1 (D) 10 [1]

3 Which of these is rational?

(A) $\sqrt{21}$ (B) $\sqrt{23}$ (C) $\sqrt{25}$ (D) $\sqrt{27}$ [1]

4 A ring bought for $2500 is sold for $3000. What is the percentage profit?

(A) 20% (B) 25% (C) 30% (D) $16\frac{2}{3}\%$ [1]

5 Consider the scores 7, 5, 2, 9, 7. Which is **not** equal to 7?

(A) mean (B) median (C) mode (D) range [1]

6 How long is it from 5:25 pm Tuesday to 9:05 am Wednesday?

(A) 15 h 40 min (B) 15 h 50 min (C) 16 h 40 min (D) 16 h 50 min [1]

7 $9 - 2(x + 3) =$

(A) $12 - 2x$ (B) $15 - 2x$ (C) $-2x - 3$ (D) $3 - 2x$ [1]

8 Which rectangular prism has the greatest volume?

(A) 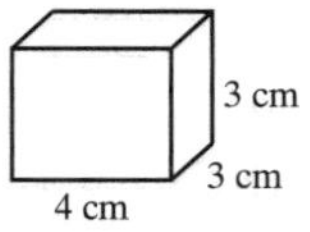

(B) 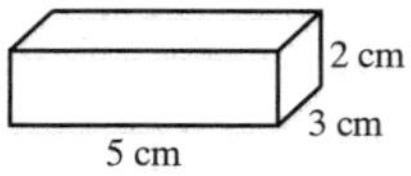

(C) 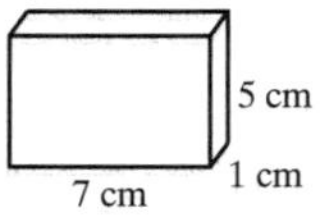

(D) 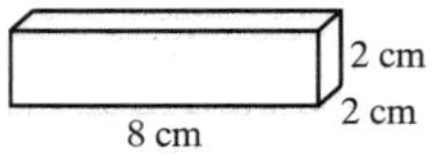

[1]

9 $-2x + 3y - 5x - 8y =$

(A) $3x - 5y$ (B) $3x - 11y$ (C) $-7x - 5y$ (D) $-7x - 11y$ [1]

10 A card is chosen from a standard pack of playing cards. What is the probability that it is red or black?

(A) $\frac{1}{2}$ (B) $\frac{3}{4}$ (C) $\frac{10}{13}$ (D) 1 [1]

11 $\frac{5}{6} =$

(A) $0.5\dot{6}$ (B) $0.8\dot{3}$ (C) $0.8\dot{6}$ (D) $0.\dot{8}$ [1]

12 What is the circumference of a circle of radius 3 cm?

(A) 6 cm (B) 3π cm (C) 6π cm (D) 9π cm [1]

EXAM PAPER 2: LEVEL 2 — PART A

Marks

13 What is the result if \$200 is increased by 150%?

(A) \$300 (B) \$350 (C) \$450 (D) \$500 — 1

14 $-0.9 \div -0.3 =$

(A) 3 (B) −3 (C) 0.3 (D) −0.3 — 1

15 What is the area of this rhombus?

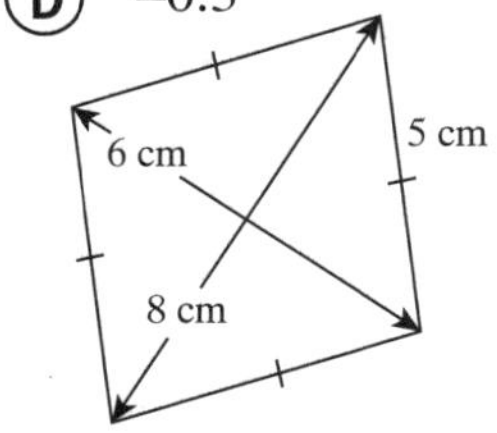

(A) 24 cm² (B) 30 cm² (C) 40 cm² (D) 48 cm² — 1

16 If Mitchell, Josh and Nathan share \$1800 in the ratio 2 : 3 : 4, what is Nathan's share?

(A) \$400 (B) \$600 (C) \$700 (D) \$800 — 1

17 $12abc^2 \div 3ab^2 =$

(A) $\frac{4b}{c^2}$ (B) $\frac{4c^2}{b}$ (C) $\frac{c^2}{4b}$ (D) $\frac{b}{4c^2}$ — 1

18 $-\frac{2}{3} \div \frac{1}{2} =$

(A) $-\frac{1}{3}$ (B) $\frac{1}{3}$ (C) $-1\frac{1}{3}$ (D) $1\frac{1}{3}$ — 1

19 Which of these is fully factorised?

(A) $12x + 18 = 3(4x + 6)$ (B) $2x^2 - 14x = x(2x - 14)$

(C) $5xy + 10x = 5(xy + x)$ (D) $12x - 9y = 3(4x - 3y)$ — 1

20 Which of these points will lie on the line $y = 3x - 5$?

(A) (1, 2) (B) (−1, −2) (C) (−2, −1) (D) (2, 1) — 1

21 What is the value of x?

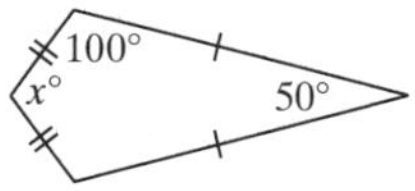

(A) 90 (B) 100 (C) 110 (D) 130 — 1

22 A car uses 7.5 L of petrol for every 100 km travelled. How much fuel is needed to travel 600 km?

(A) 42.5 L (B) 45 L (C) 48 L (D) 54 L — 1

23 Which equation has solution $x = 4$?

(A) $3x - 5 = 17$ (B) $31 - 5x = 11$ (C) $\frac{x}{2} + 7 = 15$ (D) $\frac{x+6}{2} = 7$ — 1

24 The ratio of red to blue to green counters is 5 : 3 : 2. What fraction of the counters is blue?

(A) $\frac{3}{7}$ (B) $\frac{3}{10}$ (C) $\frac{1}{3}$ (D) $\frac{1}{4}$ — 1

25 Which of these triangles is **not** necessarily congruent to ΔPQR?

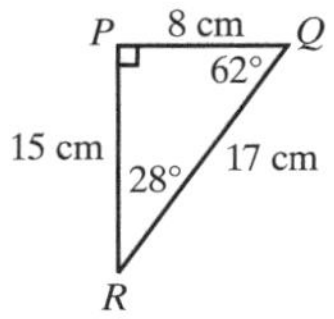

(A)
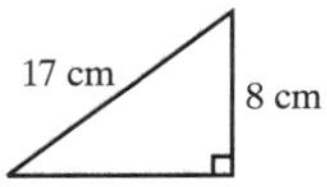

(B)
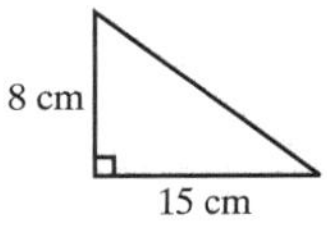

(C)
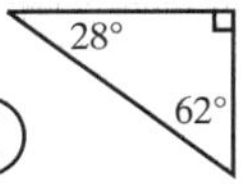

(D)
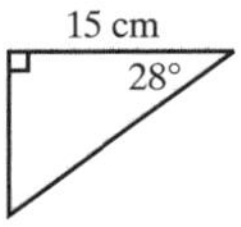

— 1

Total marks /25

EXAM PAPER 2 : LEVEL 2 PART B

Marks

1 $-2 \times -5 \times 7 \times 3 \times 0 \times -4 =$ _______________ 1

2 If $p = -2$, what is the value of $10p^2$? _______________ 1

3 How many square metres are in a square kilometre? _______________ 1

4 $(7^2)^3$ in simplest index form = _______________ 1

5 $8a \times 3ab =$ _______________ 1

6 What is the percentage increase if \$200 increases by \$500? _______________ 1

7 Factorise $7xy - 7$. _______________ 1

8 What is the probability of getting at least 4 when a dice is rolled? _______________ 1

9 Is π rational or irrational? _______________ 1

10 What is the ratio of 70 g to 2 kg in simplest form? _______________ 1

11 What is the value of x? _______________ 1

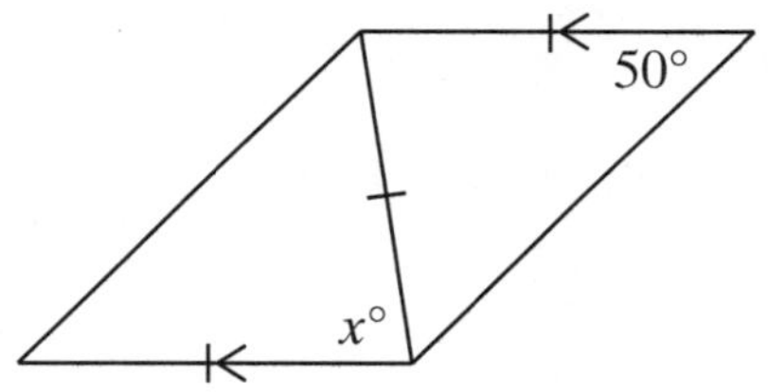

12 $5^4 \times 5^6$ in simplest index form = _______________ 1

13 What is the time $2\frac{3}{4}$ hours after 4:50 pm? _______________ 1

14 What is the missing value in this table of values? _______________ 1

x	0	1	2	3	4
y	3	7	11		19

15 $\sqrt{3^2 + 4^2} =$ _______________ 1

Marks

16 Expand $-2x(x - 5)$. ______ 1

17 $6 - 3 \times -5 =$ ______ 1

18 $\frac{2}{3} - \frac{5}{6} =$ ______ 1

19 How much is a discount of 40% on a chair originally priced at $600? ______ 1

20 What is $0.2\dot{8}$ correct to three decimal places? ______ 1

21 What is the solution to $\frac{x}{5} + 3 = 7$? ______ 1

22 If the mean of 10 scores is 17, what is the sum of the scores? ______ 1

23 Simplify $\frac{8a^2}{2ab}$. ______ 1

24 $2 - 12 \div -2 =$ ______ 1

25 What is the area of this trapezium? ______ 1

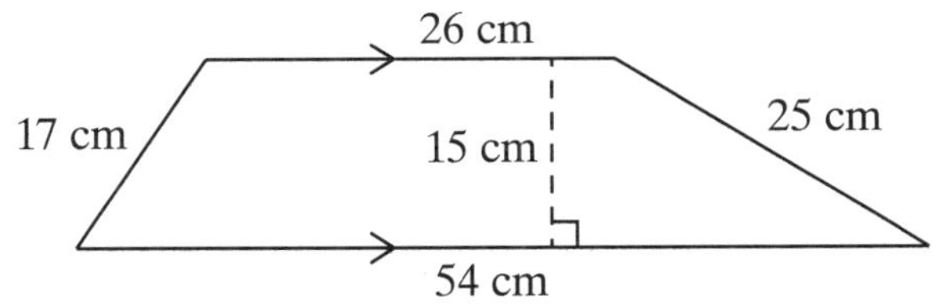

26 The ratio of cars to trucks is 5 : 2. If there are 30 trucks, how many cars are there? ______ 1

27 $-\frac{3}{4} \times -\frac{1}{2} =$ ______ 1

28 Factorise fully: $6x^2 - 9x$. ______ 1

29 $3^{15} \div 3^5$ in simplest index form = ______ 1

30 What is the average speed in kilometres per hour if a train travels 60 km in 45 minutes? ______ 1

Total marks /30

1 Given that $68 \times 73 = 4964$, find the value of:

a $-4964 \div -73$ — 1 mark

b -34×73 — 1 mark

2 Expand and simplify:

a $3(2x + 5y) - 4(x - 3y)$ — 2 marks

b $x(x - 7) - 2(x + 7)$ — 2 marks

3 A group of 60 students were asked if they had ever travelled overseas or interstate. The Venn diagram shows the result.

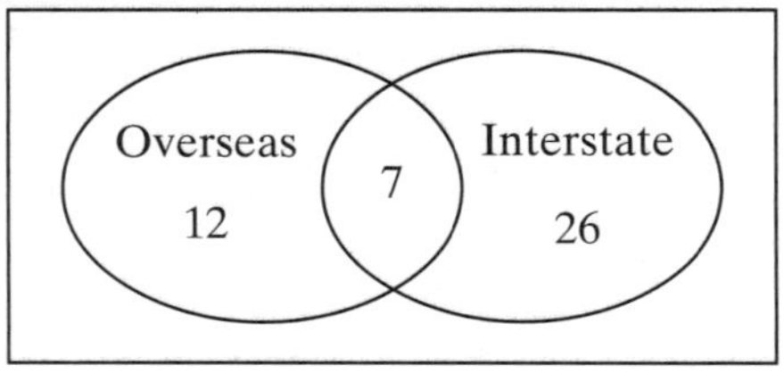

a The number of students who had neither travelled overseas nor interstate is missing. What should it be? — 1 mark

b What is the probability that a student, chosen at random from the group, has travelled interstate? — 1 mark

c Based on these results, how many of 200 students would have travelled overseas but not interstate? — 2 marks

4 Solve:

a $7x + 1 = 3x + 13$ — 2 marks

b $\frac{2x}{3} - 8 = -2$ — 2 marks

5 Express $\frac{3}{11}$ as a recurring decimal. — 2 marks

6 Emil charges \$80 per hour to repair appliances.

a How much will Emil charge for work that takes him 4 hours? — 1 mark

b GST of 10% must then be added to the amount. What is the total amount that needs to be paid? — 1 mark

c If Emil increases the amount he charges by 5%, what is the new hourly rate (before GST)? — 1 mark

7 Each of the 80 children in a kindergarten was asked their favourite colour. The results are shown in the pie chart.

Favourite colour

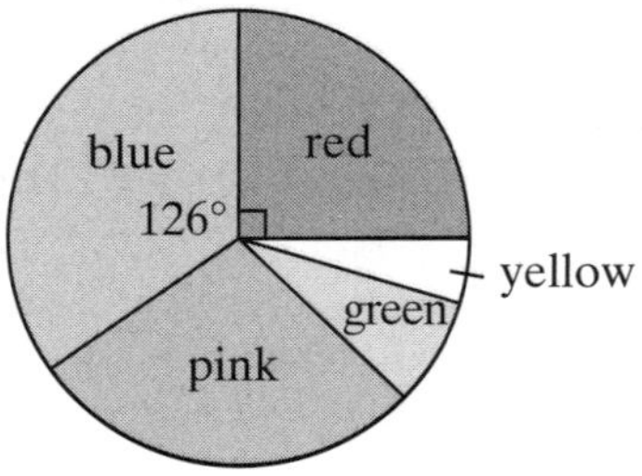

a What type of survey of the kindergarten children is this? — 1 mark

b What type of data is collected? — 1 mark

c What percentage of the children chose red as their favourite? — 1 mark

d If 22 children chose pink, what is the size of the angle in the pie chart? 1 mark

e What is the probability, as a fraction in simplest form, that a student chose blue? 1 mark

f The ratio of students who chose green to those who chose yellow is 3 : 2. How many chose green? 2 marks

8 A house was bought for $400 000 and later sold for half a million dollars. What was the percentage profit? 2 marks

9 Find $4^4 \times 8^2$ in simplest index form. 2 marks

10 The ratio of ducks to geese around a pond was 7 : 2. There were 45 ducks and geese altogether.

a How many ducks were there? 1 mark

b An extra seven ducks and four geese arrived. What is the new ratio of ducks to geese? 2 marks

11 These are the times, in minutes, some people took to complete a puzzle:
3, 3, 4, 4, 4, 5, 5, 7, 8, 17

a What is the range? 1 mark

b What is the mode? 1 mark

c What is the median? 1 mark

d What is the mean? 1 mark

e What is the new mean if the time of 17 minutes is ignored? Give the answer in fractional form. 1 mark

f Briefly explain the effect on the range, mode and median if the score of 17 is ignored. 2 marks

12 The diagram represents a paddock.

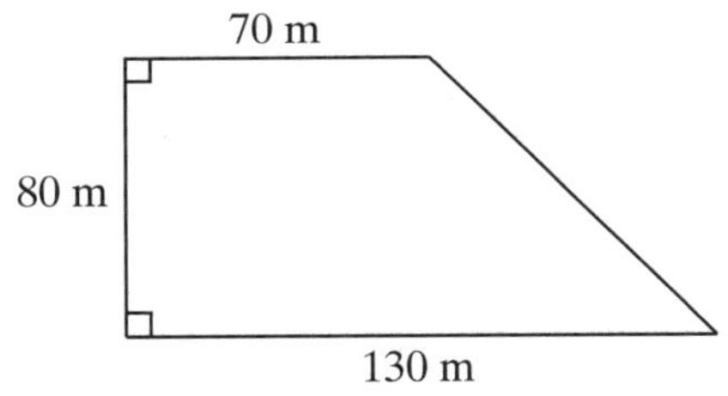

a What type of quadrilateral is the paddock? Justify your answer. 1 mark

b What is the area of the paddock in hectares? 2 marks

c What is the perimeter of the paddock? 2 marks

Total marks for PART C /45

Total marks /100

Exam Paper 3: Level 3 Non-calculator

Instructions for all parts • Attempt all questions.
Time allowed: 1 hour 40 minutes • Calculators are not allowed.

Part A: Allow about 25 minutes for this part.
Part B: Allow about 30 minutes for this part.
Part C: Allow about 45 minutes for this part. Show all working for this part. **Total marks: 100**

EXAM PAPER 3 : LEVEL 3 PART A

Fill in only one circle for each question.

Marks

1 $-90 \div 3 \times -2 + 1 =$

(A) 61 (B) 16 (C) −59 (D) −14 — 1

2 If $a = -3$ and $b = -10$ then $ab^2 =$

(A) −300 (B) 300 (C) −900 (D) 900 — 1

3 A kite has diagonals of length 9 cm and 13 cm. Which is closest to the area of the kite?

(A) 50 cm² (B) 60 cm² (C) 110 cm² (D) 120 cm² — 1

4 $7 - 3(2x - 5) =$

(A) $22 - 6x$ (B) $-6x - 8$ (C) $8x - 20$ (D) $8x + 20$ — 1

5 What is the equation of the line that passes through (−1, −5), (2, 7) and (6, 23)?

(A) $y = 2x + 1$ (B) $y = 3x + 5$ (C) $y = 4x - 1$ (D) $y = 11 - 2x$ — 1

6 A truck was bought for \$24 000 and sold for \$16 000. What was the percentage loss?

(A) 25% (B) $33\frac{1}{3}\%$ (C) 50% (D) $66\frac{2}{3}\%$ — 1

7 $(3^{12} \div 3^4)^2 =$

(A) 1 (B) 3^6 (C) 3^{10} (D) 3^{16} — 1

8 $-24p^2qr \div -6p \div -2p =$

(A) $2qr$ (B) $-2qr$ (C) $8p^2qr$ (D) $-8p^2qr$ — 1

9 $-5 \times 2 - 3 \times -4 =$

(A) 2 (B) −2 (C) −20 (D) −22 — 1

10 What is the value of x?

(A) 30 (B) 40 (C) 50 (D) 60 — 1

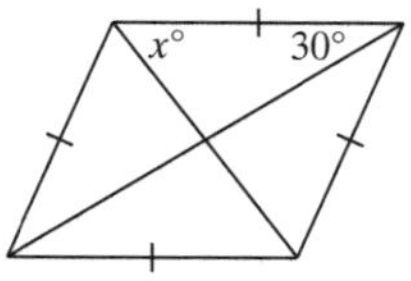

11 What is the probability of getting a number that is either odd or greater than 2, but not both, when a dice is rolled?

(A) $\frac{1}{3}$ (B) $\frac{1}{2}$ (C) $\frac{2}{3}$ (D) $\frac{5}{6}$ — 1

12 $5^0 + (3^0)^2 =$

(A) 0 (B) 1 (C) 2 (D) 3 — 1

EXAM PAPER 3 : LEVEL 3 PART A

Marks

13 How many cubic centimetres are in 0.1 m^3?

(A) 1000 (B) 10 000 (C) 100 000 (D) 1 000 000 — 1

14 Which is **not** rational?

(A) $\sqrt{4}$ (B) $\sqrt{64}$ (C) $\sqrt{124}$ (D) $\sqrt{144}$ — 1

15 Which is closest to the circumference of a circle of diameter 2 cm?

(A) 5 cm (B) 6 cm (C) 8 cm (D) 12 cm — 1

16 $\frac{1}{3} - \frac{1}{2} \times \frac{1}{4} =$

(A) $-\frac{1}{24}$ (B) $\frac{1}{24}$ (C) $-\frac{5}{24}$ (D) $\frac{5}{24}$ — 1

17 Which is the result when $8x^2 + 4x + 4$ is fully factorised?

(A) $2(4x^2 + 2x + 2)$ (B) $4(2x^2 + x)$ (C) $4(2x^2 + x + 1)$ (D) $4x(2x + 1)$ — 1

18 What is the value of x?

(A) $\sqrt{73}$ (B) $\sqrt{55}$ (C) $\sqrt{10}$ (D) $\sqrt{47}$ — 1

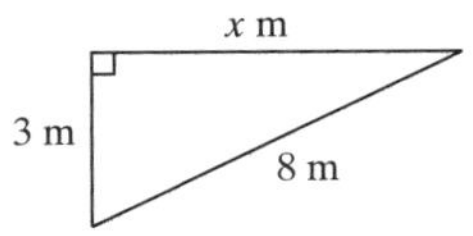

19 $4^3 \times 4^5 =$

(A) 2^{16} (B) 4^{15} (C) 16^8 (D) none of these — 1

20 A car takes $2\frac{3}{4}$ hours to travel 209 km. What is the average speed?

(A) 68 km/h (B) 72 km/h (C) 76 km/h (D) 80 km/h — 1

21 If \$800 is increased by 350% what is the result?

(A) \$1150 (B) \$2800 (C) \$3200 (D) \$3600 — 1

22 $5x^2 - 8x + 3x^2 - 11x - 4x^2 + x =$

(A) 0 (B) $4x^2 - 20x$ (C) $-2x^2 + 2x$ (D) $4x^2 - 18x$ — 1

23 What is the volume of this triangular prism?

(A) 765 cm^3 (B) 720 cm^3 (C) 408 cm^3 (D) 360 cm^3 — 1

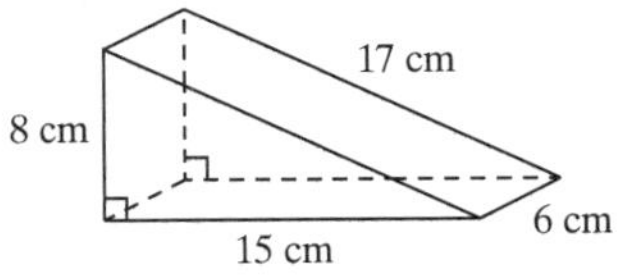

24 $0.\dot{7}\dot{2}$

(A) $\frac{8}{9}$ (B) $\frac{8}{11}$ (C) $\frac{5}{7}$ (D) $\frac{13}{18}$ — 1

25 There are red, white and blue marbles in a box. One-quarter of the marbles are red, two-fifths are blue and the rest are white. What is the ratio of red to blue to white?

(A) 5 : 8 : 7 (B) 1 : 2 : 6 (C) 1 : 2 : 7 (D) 5 : 8 : 6 — 1

Total marks /25

Marks

1 $-5 \times -17 \times -2 =$ ____ 1

2 How many litres are in 0.07 ML? ____ 1

3 What is the date 56 days after 28 September? ____ 1

4 Fill in the next number in this pattern:
1, 4, 9, 16, 25, ? ____ 1

5 Expand $-3x(5y - x - 1)$. ____ 1

6 An increase from \$8 to \$24 is what percentage increase? ____ 1

7 Write 15 cm to 6 m as a ratio in simplest form. ____ 1

8 $18a^2bc \div 3abc^2 =$ ____ 1

9 One side of a rectangle is 12 cm long. If the perimeter is 40 cm, what are the lengths of the remaining sides? ____ 1

10 $2^3 \times 4^2$ in simplest index form = ____ 1

11 If $a = -3$ and $b = -5$, what is $b^2 - 3ab$? ____ 1

12 How far will a vehicle travel in 45 minutes at a speed of 90 km/h? ____ 1

13 What is the mean of 1, 2, 3, 3, 4, 5, 6, 7, 7, 8 and 9? ____ 1

14 When expressed as a decimal, does $\frac{5}{12}$ terminate or repeat? ____ 1

15 $\frac{-7 - 5}{9 - -3} =$ ____ 1

16 If \$700 is divided in the ratio 3 : 2 : 5, what is the smallest amount? ____ 1

17 Solve $\frac{2x + 5}{3} = 4$. ____ 1

18 The numbers from 1 to 20, inclusive, are written on separate cards. If one card is selected at random, what is the probability that the number is at least 16? ____ 1

EXAM PAPER 3 : LEVEL 3 — PART B

Marks

19 Factorise $-8pq - 4q$ fully. ______ [1]

20 $-\frac{3}{5} \times 1\frac{2}{3} =$ ______ [1]

21 $ABCD$ is a rhombus. If $\angle BAD = 140°$, what is the size of $\angle BDC$? ______ [1]

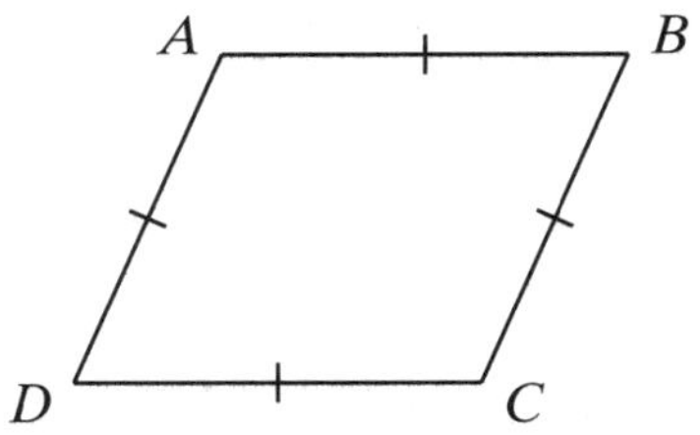

22 $37 \times 84 - 16 \times -37 =$ ______ [1]

23 $5^7 \times 5^8 \div 5^3$ in simplest index form = ______ [1]

24 What is the rule for the values in the table? ______ [1]

x	1	2	3	4
y	−2	1	4	7

25 A cylinder of height 3 cm has radius 2 cm. What is its volume in terms of π? ______ [1]

26 $-2p^2 + 5q - 3p^2 + 6q + 12p - 4p^2 =$ ______ [1]

27 What is the value of $2^5 \times 5^6$? ______ [1]

28 What is the capacity, in litres, of a triangular prism with cross-sectional area 20 cm^2 and perpendicular height 5 cm? ______ [1]

29 What is $\frac{1}{6}$ as a decimal correct to two decimal places? ______ [1]

30 A table was sold for $1500 which was a 25% profit on the cost price. What was the cost price? ______ [1]

Total marks /30

EXAM PAPER 3 : LEVEL 3 — PART C

1 Given that $11 \times 9 \times 7 = 693$, find the value of:

a $-11 \times -9 \times 7 \times -10$ [1 mark]

b $-693 \div -99$ [1 mark]

2 A shop offers 30% off all marked prices.

a How much would you save on a toy marked \$25? [1 mark]

b How much would you pay for a game marked \$39? [2 marks]

c How much was saved on a puzzle bought for \$35? [2 marks]

3 **a** Briefly explain why $2\frac{7}{11}$ is a rational number. [1 mark]

b Express $2\frac{7}{11}$ as a decimal. [2 marks]

4 In a group of 40 friends, 26 play golf and 18 play tennis, but seven play neither.

a Complete the Venn diagram showing this information. [2 marks]

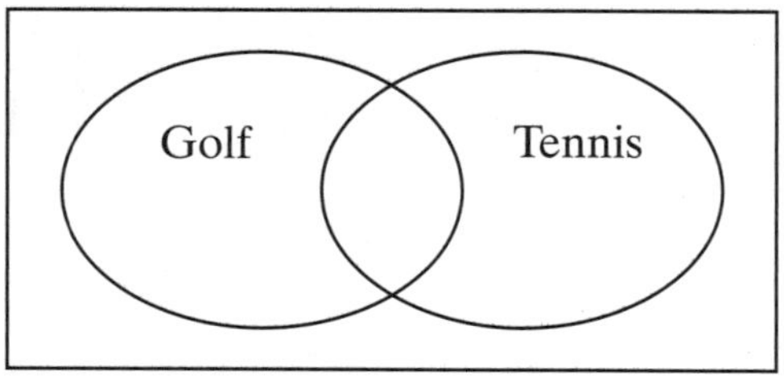

b What is the probability that a member of the group plays both golf and tennis? [1 mark]

c What is the probability that one of the friends plays at most one of the sports? [1 mark]

5 Write in simplest index form:

a $\dfrac{4^9}{16^3}$ [2 marks]

b $\dfrac{\left((3^5)^2 \times 3^4\right)^2}{(3^8 \div 3^2)^3}$ [2 marks]

6 *PQRS* is a quadrilateral with *PQ* parallel to *SR*.

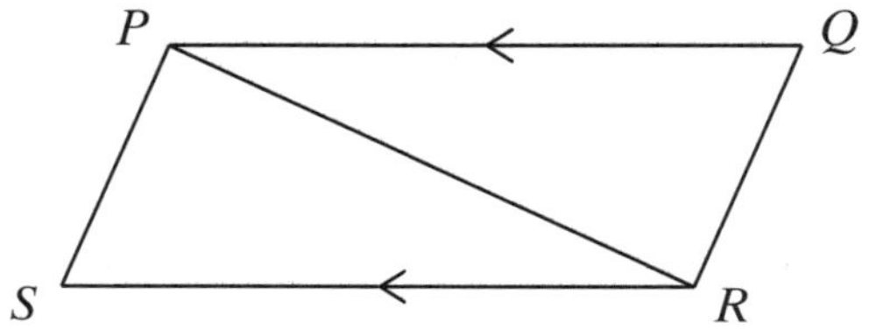

a Why does $\angle QPR = \angle SRP$? [1 mark]

b If $PQ = SR$, briefly explain why triangles *PQR* and *SRP* must be congruent. [2 marks]

c Briefly explain how this shows that the opposite angles of a parallelogram must be equal. [2 marks]

7 Solve:

a $8x + 7 = 2x - 2$ [2 marks]

b $\frac{3x}{5} - 6 = -11$ [2 marks]

8 Milk powder fed to calves is to be used at the rate of 1.2 kg in 10 L of water.

a What is the rate in grams per litre? [1 mark]

b How much milk powder would be needed with 7.5 L of water? [1 mark]

c Mel buys a 20 kg pack of the powder to feed a calf. The calf is fed twice a day. If the powder is fed in 7.5 L of water each time, for how many days would Mel have enough feed? [2 marks]

9 One-third of the tickets in a raffle are green and two-fifths are white. The rest of the tickets are blue.

a What fraction of the tickets are blue? [1 mark]

b What is the ratio of white to blue to green tickets in simplest form? [1 mark]

c If there are 80 blue tickets, how many are white and how many are green? [2 marks]

10 A test was given to two classes. The 20 students in 8H had a mean mark of 80 and the 12 students in 8K had a mean of 90. Find the mean mark for the test. [2 marks]

11 Using $\frac{22}{7}$ as an approximate value of π, what is the area of a circle of radius 3.5 cm? [2 marks]

12 $ABCD$ is a rhombus. The diagonals meet at X. $AC = 6$ cm and $BD = 8$ cm.

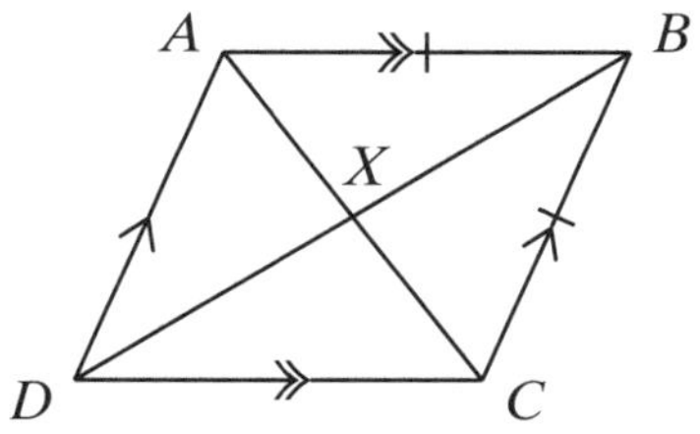

a What is the area of the rhombus? [1 mark]

b Explain why $\angle DXC = 90°$ and why it is possible to find the lengths of XD and XC. [1 mark]

c Find the length of DC. [2 marks]

d Find the perpendicular height of the rhombus. [2 marks]

Total marks for PART C /45 **Total marks** /100

Exam Paper 4: Level 1 Calculator Allowed

Instructions for all parts • Attempt all questions.
Time allowed: 1 hour 40 minutes • Calculators are allowed.

Part A: Allow about 25 minutes for this part.
Part B: Allow about 30 minutes for this part.
Part C: Allow about 45 minutes for this part. Show all working for this part. **Total marks: 100**

EXAM PAPER 4: LEVEL 1 PART A

Fill in only one circle for each question.

Marks

1 Which is a common factor of $12x^2$ and $15xy$?
(A) $2x^2$ (B) $3x$ (C) $4xy$ (D) $5y$ [1]

2 $39 \div -3 \times (-5 - -8) =$
(A) -39 (B) $-4\frac{1}{3}$ (C) 169 (D) -169 [1]

3 How many square millimetres are in 9 cm^2?
(A) 30 (B) 90 (C) 300 (D) 900 [1]

4 A movie begins at 11:25 am and runs for $2\frac{3}{4}$ hours. When does it finish?
(A) 1 pm (B) 1:10 pm (C) 2 pm (D) 2:10 pm [1]

5 Which test proves these triangles congruent?
(A) AAS (B) RHS (C) SAS (D) SSS [1]

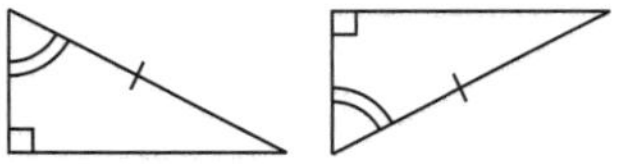

6 What is the result if \$70 000 is increased by 375%?
(A) \$192 500 (B) \$2 625 000 (C) \$262 500 (D) \$332 500 [1]

7 $5(3x - 2) =$
(A) $8x - 7$ (B) $8x - 10$ (C) $15x - 7$ (D) $15x - 10$ [1]

8 Which of these is closest to the circumference of a circle of radius 23 cm?
(A) 72 cm (B) 145 cm (C) 415 cm (D) 1662 cm [1]

9 What would be an outlier in this data set? 37, 31, 28, 26, 33, 10, 32, 24, 31, 40, 35, 30
(A) 37 (B) 10 (C) 40 (D) 30 [1]

10 Under which of these transformations is an image **not** congruent to the original shape?
(A) enlargement (B) rotation (C) reflection (D) translation [1]

11 If $a = -3$, what is the value of $2a^2$?
(A) -36 (B) 36 (C) -18 (D) 18 [1]

12 Travelling 2163 km in $5\frac{1}{4}$ hours is what average speed?
(A) 388 km/h (B) 400 km/h (C) 412 km/h (D) 420 km/h [1]

13 $5y + 6z - 8y + 3z =$
(A) $13y + 9z$ (B) $13y + 3z$ (C) $9z - 3y$ (D) $3z - 3y$ [1]

Marks

14 Which shows the graph of $y = 3x - 2$?

(A) 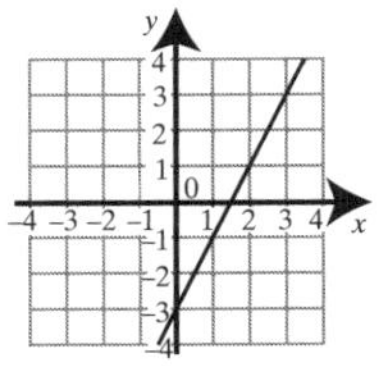

(B) 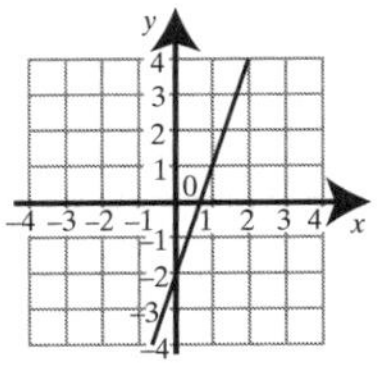

(C) 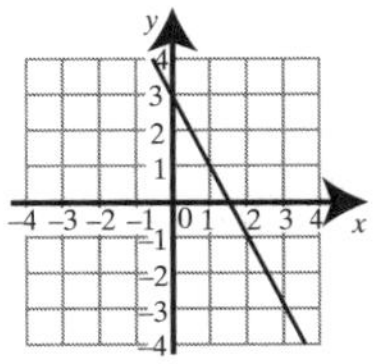

(D) 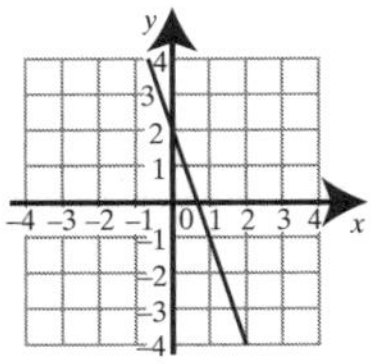

1

15 What is the ratio of boys to girls in simplest form if, out of 180 children, 75 are girls?

(A) 7 : 5 (B) 5 : 7 (C) 5 : 12 (D) 7 : 12 — 1

16 $6 - -24 \div -3 =$

(A) -2 (B) -10 (C) 6 (D) 14 — 1

17 What is the area of this rhombus in square centimetres?

(A) 324.24 (B) 319.2 (C) 638.4 (D) 372.49 — 1

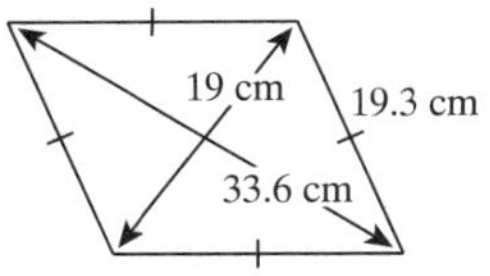

18 What is $\frac{19}{55}$ as a decimal?

(A) 0.345 (B) $0.34\dot{5}$ (C) $0.3\dot{4}\dot{5}$ (D) $0.\dot{3}\dot{4}\dot{5}$ — 1

19 If $\frac{3x}{4} - 5 = 7$ then $x =$

(A) $2\frac{2}{3}$ (B) $1\frac{1}{2}$ (C) 9 (D) 16 — 1

20 A box holds marbles in three different colours. Twenty-two of the marbles are blue, sixteen are green and eighteen yellow. If one marble is taken from the box without looking, what is the probability that it is green?

(A) $\frac{1}{3}$ (B) $\frac{2}{5}$ (C) $\frac{2}{7}$ (D) $\frac{1}{4}$ — 1

21 $24x \div 2x \times 3x =$

(A) $36x$ (B) $4x$ (C) $\frac{4}{x}$ (D) $36x^2$ — 1

22 What is the value of x?

(A) 98 (B) 178 (C) 224 (D) 369 — 1

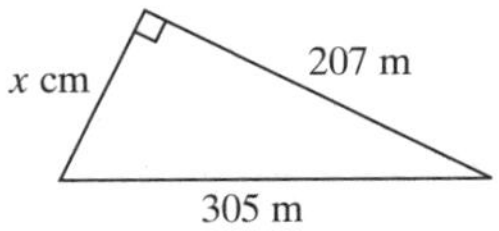

23 $2^7 \times 3^7 =$

(A) 5^7 (B) 6^7 (C) 5^{14} (D) 6^{14} — 1

24 What is the selling price, if an item was bought for \$900 and sold at a loss of 16%?

(A) \$144 (B) \$740 (C) \$756 (D) \$843.75 — 1

25 What is the volume of this prism?

(A) 1680 cm^3 (B) 1785 cm^3 (C) 1904 cm^3 (D) 2040 cm^3 — 1

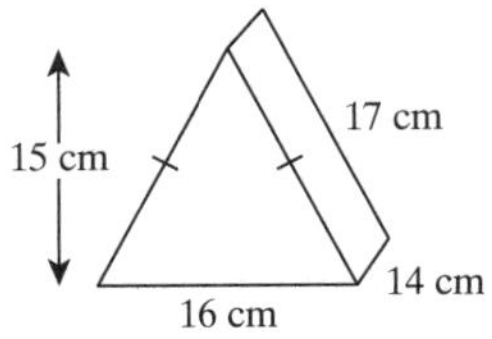

Total marks /25

EXAM PAPER 4: LEVEL 1 — PART B

Marks

1 What is the median of the scores 7, 10, 12, 8, 10 and 5? ____ 1

2 What is the probability of **not** getting a 6 when a dice is rolled? ____ 1

3 Factorise $27a + 45b$. ____ 1

4 What is the area of a parallelogram with base 17 cm and perpendicular height 9 cm? ____ 1

5 What is $\sqrt{42}$ to one decimal place? ____ 1

6 $7x^2 + 3x + 5x =$ ____ 1

7 $-24 + 8 \times -3 - -6 \times -2 =$ ____ 1

8 What is the missing value in this table? ____ 1

x	0	1	2	3	4
y	14	11		5	2

9 What is the ratio 18 : 24 : 27 in simplest form? ____ 1

10 If $p = -3$ and $q = -4$, what is the value of $p - pq$? ____ 1

11 Decrease \$1500 by 15%. ____ 1

12 What is the solution to $17 - 3x = 2$? ____ 1

13 $\dfrac{-21 \div -7}{3 + 2 \times -5}$ as a simple fraction = ____ 1

14 How many days is 432 hours? ____ 1

15 Expand $7(3x - 8)$. ____ 1

16 $56mn \div 8n =$ ____ 1

17 What is $\dfrac{3}{16}$ as a decimal? ____ 1

18 What is the profit if a house is bought for \$860 000 and sold for \$1.2 million? ____ 1

19 What is the diameter, in metres, of a circle of radius 85 cm? ____ 1

20 $6a^2 + 4a \times 3a =$ ____ 1

EXAM PAPER 4: LEVEL 1 — PART B

Marks

21 $-2\frac{3}{4} \div -1\frac{1}{2}$ as a mixed number = ______ [1]

22 If, in 20 seconds, 3 L of water flows, what is the rate in litres per minute? ______ [1]

23 Referring to the graph below, what is the point of intersection of the lines $y = x - 1$ and $y = 1 - x$? ______ [1]

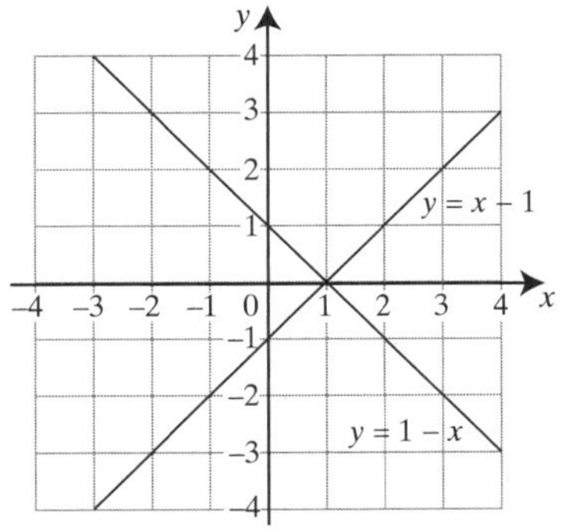

24 If $\Delta ABC \equiv \Delta YXZ$, which side of ΔXYZ corresponds to AC of ΔABC? ______ [1]

25 What is $0.2\dot{5}$ as a fraction in simplest form? ______ [1]

26 What is $7^8 \div 7^2$ in index form? ______ [1]

27 The area of the constant cross-section of a hexagonal prism is 14.2 m^2 and the perpendicular height is 3.7 m. What is the volume? ______ [1]

28 If Bill and Ben share $756 in the ratio 2 : 7, what is Ben's share? ______ [1]

29 What is the length of each side of a square with the same perimeter as this rectangle? ______ [1]

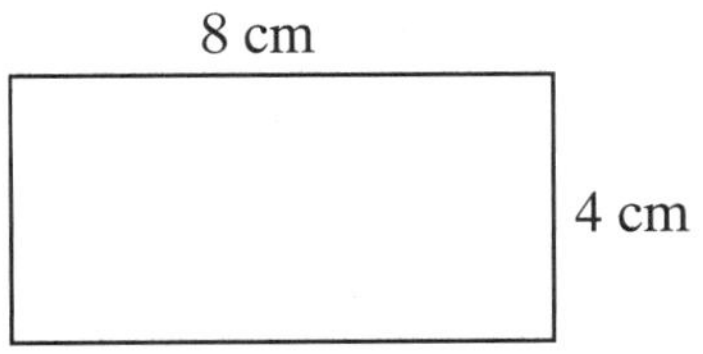

30 $5 \times 5^0 =$ ______ [1]

Total marks /30

1 Expand and simplify $5(7x - 8y) - 6(9x - 4y)$. [2 marks]

2 Factorise fully: $15a^2b^2 - 10a^2b$. [2 marks]

3 Tom ran 1500 m in 4 minutes.

a At this rate, how far did Tom run in 1 minute? [1 mark]

b How far, in metres, could Tom run in an hour at the same rate? [1 mark]

c What was Tom's average speed in kilometres per hour? [1 mark]

4 **a** Is $\frac{3}{7}$ rational or irrational? Justify your answer. [1 mark]

b What is $\frac{3}{7}$ as a decimal? [1 mark]

c What is $\frac{3}{7}$ as a decimal rounded to three decimal places? [1 mark]

5 A circle of diameter 14 cm fits exactly in a square as shown in the diagram.

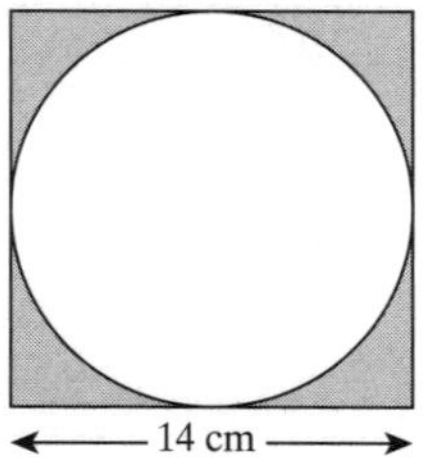

a What is the area of the circle? Give the answer to two decimal places. [2 marks]

b What is the area of the square? [1 mark]

c What is the shaded area to the nearest square centimetre? [1 mark]

6 A group of 270 people were asked if they were right- or left-handed. The results are shown in the table.

	Left-handed	Right-handed
Male	19	123
Female	17	

a The number of right-handed females is missing. What should it be? [1 mark]

b What is the probability that a member of the group is left-handed? [1 mark]

c What is the probability that a member of the group is a right-handed male? [1 mark]

d What is the probability that a member of the group is male or left-handed or both? [1 mark]

7 Solve:

a $9x + 17 = 33 - x$ [2 marks]

b $\frac{2x - 5}{9} = -7$ [2 marks]

8 A stall-holder at a market made a total of \$2520 last week. That gave her a profit of \$270.

a What were her costs? [1 mark]

b What was the profit as a percentage of the costs? 1 mark

c The next week her profit increased by 40%. If her costs remained the same, how much money did she make in total in the next week? 2 marks

9 a Complete the table for $y = -2x + 7$. 2 marks

x	0	1	2	3
y				

b Plot the points from the table and graph the line $y = -2x + 7$ on the number plane. 2 marks

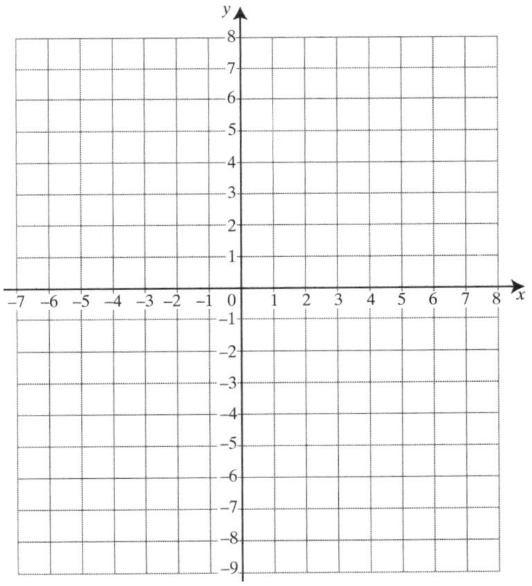

c What is the solution to $-2x + 7 = -9$. 1 mark

10 The ages of the students in a course at a community college were:
29, 36, 24, 35, 41, 24, 27, 31, 28, 22, 33.

a What was the range? 1 mark

b What was the mode? 1 mark

c What was the median? 1 mark

d What was the mean? 1 mark

e Another student, aged 30, joined the course. What is the mean now? Justify your answer. 1 mark

f Is this group of students a good representation of the students attending the college? Justify your answer. 2 marks

11 a What is the length of the third side of this triangle? 2 marks

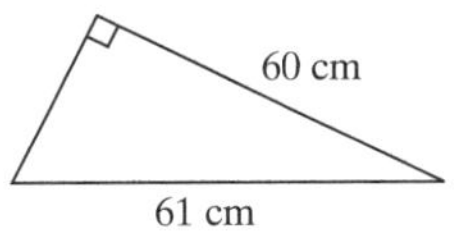

b What is the area of the triangle? 1 mark

c If the triangle is the cross-section of a prism of height 20 cm, what is the volume? 1 mark

d What is the capacity of the prism in litres? 1 mark

e How many fewer litres will this prism hold than a cube of side length 20 cm? 2 marks

Total marks for PART C /45 **Total marks** /100

Exam Paper 5: Level 2 Calculator Allowed

Instructions for all parts • Attempt all questions.
Time allowed: 1 hour 40 minutes • Calculators are allowed.

Part A: Allow about 25 minutes for this part.
Part B: Allow about 30 minutes for this part.
Part C: Allow about 45 minutes for this part. Show all working for this part. **Total marks: 100**

EXAM PAPER 5: LEVEL 2 — PART A

Fill in only one circle for each question.

Marks

1 $-5 + 6 \times -4 =$

(A) -29 (B) -4 (C) 19 (D) -19 — 1

2 $x(x + 9) =$

(A) $2x + 9$ (B) $9x^2$ (C) $11x$ (D) $x^2 + 9x$ — 1

3 x is closest to:

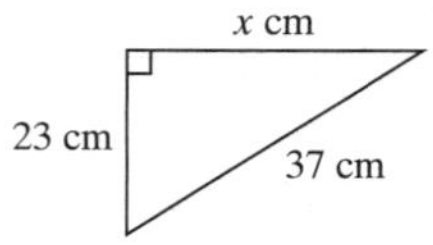

(A) 27 (B) 29 (C) 35 (D) 44 — 1

4 The price of a house increased by 15%. If the old price was \$680 000 what was the new price?

(A) \$690 200 (B) \$771 200 (C) \$782 000 (D) \$830 000 — 1

5 How long would it take to travel 240 km at 75 km/h?

(A) 3 h 2 min (B) 3 h 12 min (C) 3 h 20 min (D) 3 h 40 min — 1

6 What is the area of this parallelogram?

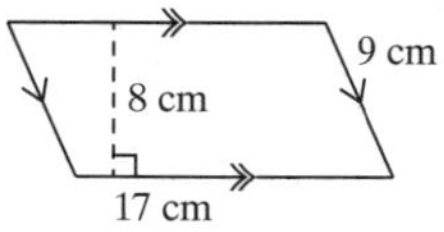

(A) 72 cm² (B) 136 cm² (C) 153 cm² (D) 289 cm² — 1

7 What is the ratio $\frac{1}{3} : \frac{1}{2} : \frac{1}{4}$ in simplest form?

(A) 3 : 2 : 4 (B) 6 : 7 : 5 (C) 6 : 4 : 3 (D) 4 : 6 : 3 — 1

8 Which is $28x^2y - 35xy^2$ fully factorised?

(A) $7(4x^2y - 5xy^2)$ (B) $xy(28x - 35y)$ (C) $7xy^2(4x - 5)$ (D) $7xy(4x - 5y)$ — 1

9 What is the circumference, to the nearest centimetre, of a circle of radius 9 cm?

(A) 28 cm (B) 57 cm (C) 64 cm (D) 254 cm — 1

10 These cards are placed in a hat. If one card is chosen at random, what is the probability that the number it shows is at least 5?

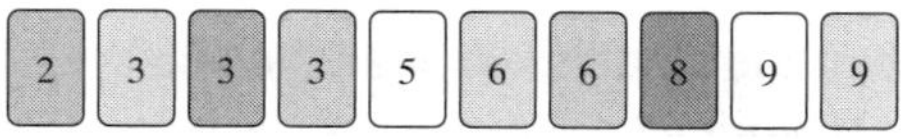

(A) $\frac{1}{2}$ (B) $\frac{2}{5}$ (C) $\frac{3}{5}$ (D) $\frac{1}{10}$ — 1

11 What is $\frac{59}{66}$ as a decimal?

(A) $0.\dot{8}9\dot{3}$ (B) $0.8\dot{9}3\dot{9}$ (C) $0.89\dot{3}$ (D) $0.8\dot{9}\dot{3}$ — 1

Marks

12 $(2^5 \times 2^4)^3 =$

(A) 2^{12} (B) 2^{23} (C) 2^{27} (D) 2^{60} 1

13 $-4p^2 - 3p + 2p^2 - 9p =$

(A) $-2p^2 - 12p$ (B) $-2p^2 - 6p$ (C) $-6p^2 - 12p$ (D) $-6p^2 - 6p$ 1

14 Which is an example of categorical data?

(A) make of cars (B) speed of cars (C) length of cars (D) number of cars 1

15 What is the solution to the equation $2x - 18 = 42$?

(A) $x = 12$ (B) $x = 30$ (C) $x = 48$ (D) $x = 24$ 1

16 Multiplying an amount by 1.2 will increase it by:

(A) 2% (B) 12% (C) 20% (D) 120% 1

17 Which angle of ΔSTU corresponds to $\angle PQR$ of ΔPQR?

(A) $\angle SUT$ (B) $\angle STU$ (C) $\angle TSU$ (D) $\angle UTS$ 1

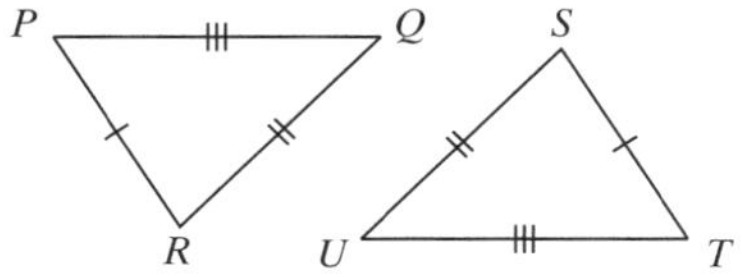

18 $\dfrac{-28 + 3 \times -8}{16 - -4 \times 9}$

(A) -1 (B) 2.6 (C) 6.5 (D) 32.75 1

19 Which of these is rational?

(A) $\sqrt{464}$ (B) $\sqrt{474}$ (C) $\sqrt{484}$ (D) $\sqrt{494}$ 1

20 How many cubic centimetres are in 0.1 m^3?

(A) 1000 (B) 10 000 (C) 100 000 (D) 1 000 000 1

21 The ratio of red to green apples is 7 : 5. If there are 84 apples altogether, how many more red than green apples are there?

(A) 12 (B) 14 (C) 24 (D) 36 1

22 What is the date 36 days before 11 November?

(A) 3 October (B) 4 October (C) 5 October (D) 6 October 1

23 If $n = -7$, what is the value of $5n^2$?

(A) 245 (B) -245 (C) 1225 (D) -1225 1

24 $x =$

(A) 40 (B) 50 (C) 90 (D) 100 1

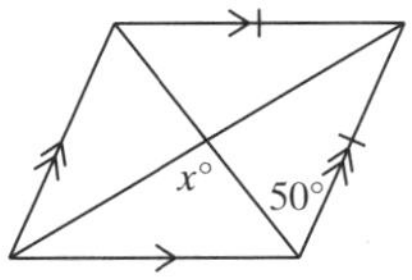

25 Which point does **not** lie on the line $y = 3x - 5$?

(A) (163, 484) (B) (4, 7) (C) (−2, −11) (D) (−1, 2) 1

Total marks /25

Marks

1 $72a^2 \div 3a \times 4a =$ ____ 1

2 What is $\frac{21}{45}$ expressed as a decimal correct to two decimal places? ____ 1

3 $-3 \times -5 \times -9 \times -8 \times -6 =$ ____ 1

4 What is the area of a square with perimeter 20 m? ____ 1

5 What is the ratio of 60 cents to $3 in simplest form? ____ 1

6 What is the percentage increase if 700 g increases to 1.4 kg? ____ 1

7 $-3\frac{1}{4} \div \frac{2}{5} =$ ____ 1

8 What is the solution to $\frac{3x}{2} = 12$? ____ 1

9 Expand $-3x(2x - 5)$. ____ 1

10 What is the volume, in cubic metres, of a rectangular prism 3 m long, 2 m wide and 80 cm high? ____ 1

11 How long is it from 11:35 am until 2:20 pm? ____ 1

12 Fully factorise $8x^2y - 4xy$. ____ 1

13 $-7x^2 + 5x - x^2 + 7x - x =$ ____ 1

14 What is $(4^2)^3$ in simplest index form? ____ 1

15 $\sqrt{41^2 - 40^2} =$ ____ 1

16 How long, in hours and minutes, does it take to travel 234 km at 72 km/h? ____ 1

17 At what point does the line $y = 2x + 4$ cut the y-axis? ____ 1

18 $\frac{-48 \times -7}{56 \div -4} =$ ____ 1

EXAM PAPER 5 : LEVEL 2 PART B

Marks

19 The mean of seven scores is 86. A new score of 95 is added. What is the new mean? ____________ 1

20 The probability of winning a lottery is $\frac{1}{100\,000}$. What is the probability of **not** winning? ____________ 1

21 *PQRS* is a kite. $\angle PQR = 110°$ and $\angle QRS = 60°$. What is the size of $\angle QPS$? ____________ 1

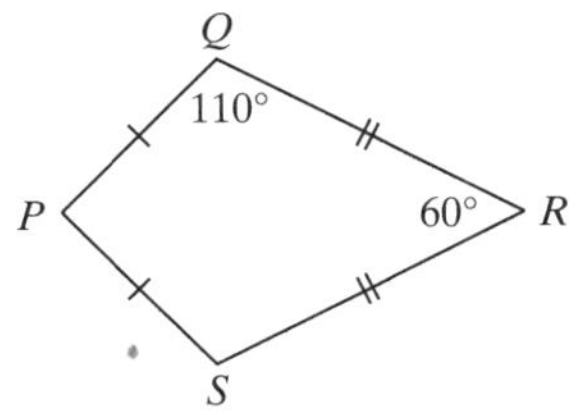

22 What special quadrilateral has all its sides equal and all its angles equal? ____________ 1

23 If $x^2 = 946\ 729$ and x is positive, what is its value? ____________ 1

24 $(-9 - 17) \times -11 + 72 \div -4 =$ ____________ 1

25 Find the percentage increase if \$1800 is increased by \$2400. ____________ 1

26 When Jai left home one morning he noticed the time on the clock, as shown. If he returned home $6\frac{3}{4}$ hours later, what was the time then? ____________ 1

27 What is the value of $12^7 + 12^0$? ____________ 1

28 If $x = -3$ and $y = 5$, what is the value of $x^2 - y^2$? ____________ 1

29 What is the radius of a circle, in centimetres, if the diameter is 0.6 m? ____________ 1

30 How many kilolitres are in 0.3 ML? ____________ 1

Total marks /30

1 Expand and simplify $5x - 2(x - 3) - 4(3x + 1)$. 2 marks

2 **a** A fridge was sold for $880. If that is a profit of 25%, what was the cost price? 2 marks

b The price included GST (10%). How much was the GST? 1 mark

3 Consider this pattern:

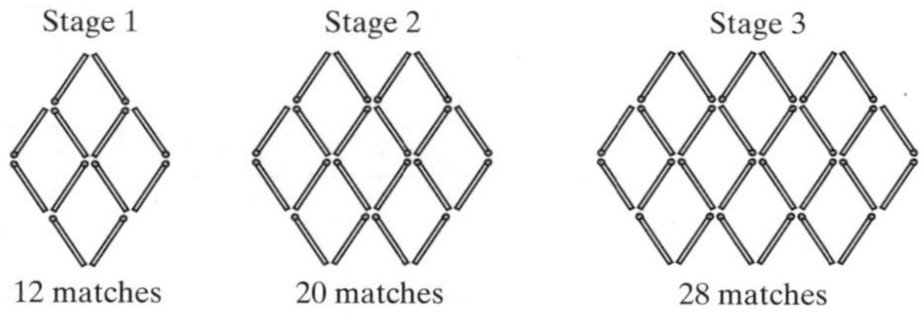

a Fill in the rest of the table. 1 mark

Stage number (s)	1	2	3	4	5	6
Number of matches (m)	12	20	28			

b Write a rule in words to explain how the number of matches is related to the stage number. 1 mark

c Write the rule using numbers and symbols. 1 mark

d How many matches would be needed at Stage 17? 1 mark

e At what stage would 236 matches be needed? 1 mark

4 Solve $7x - 13 = 26 - 4x$. 2 marks

5 These are the heights (in centimetres) of a sample of students at a school.
153, 167, 158, 170, 162, 185, 172, 166, 151, 167, 174, 163

a Is the data numerical or categorical? 1 mark

b Briefly explain how this sample would differ from a census. 1 mark

c What is the range? 1 mark

d What is the mean? 1 mark

e What is the mode? 1 mark

f What is the median? 1 mark

g Which, if any, of the mean, mode, median and range would increase and which would decrease if the height of 185 cm was left out? 3 marks

6 A swimming pool has a trapezoidal cross-section as shown.

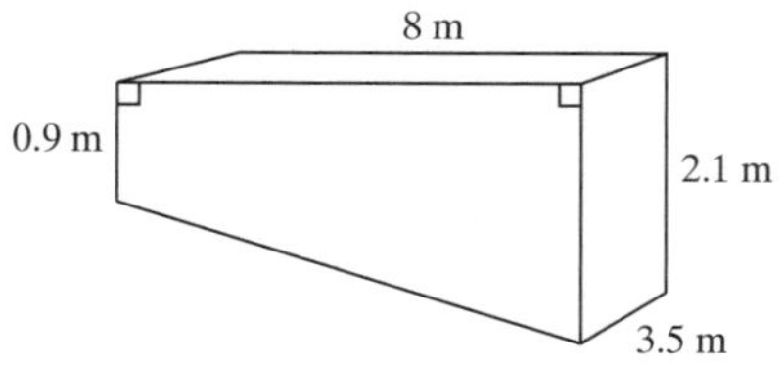

a What is the volume of the pool in cubic metres? 2 marks

b How many litres of water will the pool hold when full? 1 mark

c The pool is being filled through a hose. The water flows at a constant rate. It begins to fill at 2 pm on Saturday and is finally full at 10 am on the following Friday. At what rate, in litres per minute, did the water flow into the pool? 2 marks

7 Some students were asked whether they had ever played hockey or basketball. The results are shown in the Venn diagram.

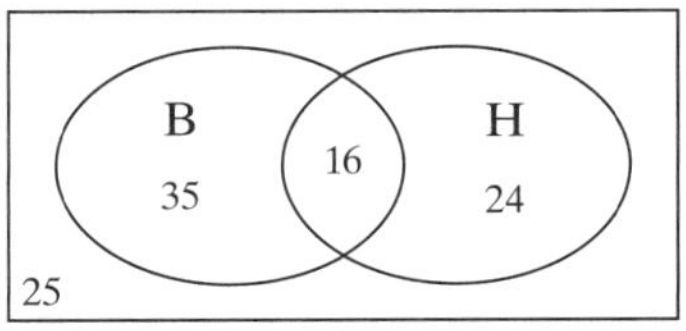

a How many students were surveyed? 1 mark

b What is the probability that a student had played hockey? 1 mark

c What is the probability that a student had played either hockey or basketball but not both? 1 mark

8 *ABCD* is a parallelogram. The diagonals meet at *P*.

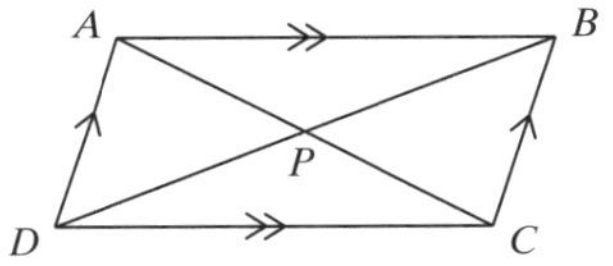

a Why does $AD = CB$? 1 mark

b Why does $\angle APD = \angle CPB$? 1 mark

c Why does $\angle DAC = \angle BCA$? 1 mark

d What conclusion can be drawn about ΔAPD and ΔCPB? 1 mark

e Why then, must $AP = CP$? 1 mark

f What special property of parallelograms does this show? 1 mark

9 There were 24 muffins on a tray. The muffins had chocolate and banana flavours in the ratio 5 : 3. Four of the muffins were eaten. The ratio of chocolate to banana muffins is now 3 : 2. How many banana muffins were eaten? 3 marks

10 The diagram shows a yard made up of a triangle and semicircle.

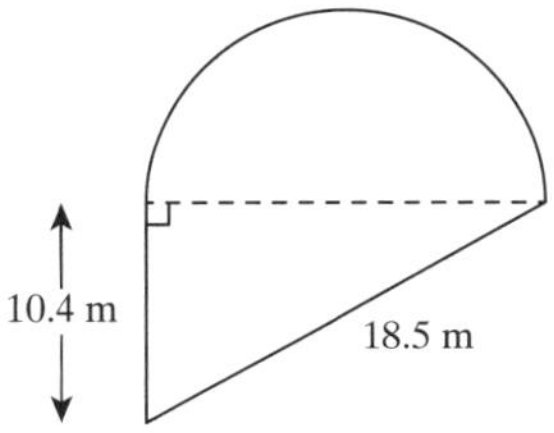

a Find the diameter of the semicircle. 2 marks

b Find the area of the yard. Give the answer in square metres correct to one decimal place. 3 marks

c Find the perimeter of the yard to the nearest metre. 2 marks

Total marks for PART C /45 **Total marks** /100

Exam Paper 6: Level 3 Calculator Allowed

Instructions for all parts • Attempt all questions.
Time allowed: 1 hour 40 minutes • Calculators are allowed.

Part A: Allow about 25 minutes for this part.
Part B: Allow about 30 minutes for this part.
Part C: Allow about 45 minutes for this part. Show all working for this part. **Total marks: 100**

EXAM PAPER 6: LEVEL 3 PART A

Fill in only one circle for each question.

Marks

1 What is $5\frac{11}{12}$ correct to three decimal places?

(A) 5.92 (B) 5.916 (C) 5.917 (D) 5.111 [1]

2 How long would it take to travel 168 km at 70 km/h?

(A) 2 h 4 min (B) 2 h 14 min (C) 2 h 24 min (D) 2h 40 min [1]

3 $(-5x - 7x) \times (-4x + 9x) =$

(A) $-60x^2$ (B) $-26x^2$ (C) $10x^2$ (D) $156x^2$ [1]

4 What is the capacity of this rectangular prism?

(A) 18 kL (B) 1800 L (C) 180 L (D) 18 L [1]

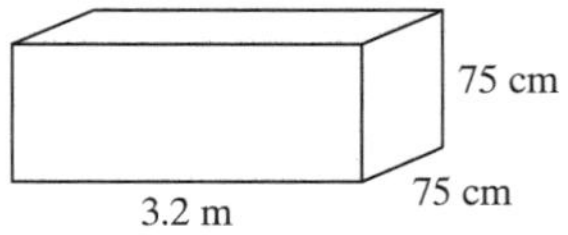

5 $3645 \div -15 \times (-8 + (-19)) =$

(A) 9 (B) 1925 (C) –2673 (D) 6561 [1]

6 How many days from, and including, 28 November until, and including, 7 January?

(A) 39 (B) 40 (C) 41 (D) 42 [1]

7 What is the ratio 85 cm to $3\frac{1}{4}$ m in simplest form?

(A) 17 : 65 (B) 65 : 17 (C) 13 : 340 (D) 340 : 13 [1]

8 $-7p - 3p(p - 1) =$

(A) $-10p^2 + 10p$ (B) $-3p^2 - 4p$ (C) $-10p - 3p^2$ (D) $-10p + 3p^2$ [1]

9 The area of a quadrant of a circle of radius 17 cm would be closest to which of these?

(A) 58 cm^2 (B) 107 cm^2 (C) 182 cm^2 (D) 227 cm^2 [1]

10 For the scores 48, 37, 41, 46, 39, 41, 45, 23, 38, 41 and 46 which of these will decrease if the score of 23 is ignored?

(A) mean (B) mode (C) median (D) range [1]

11 If $\frac{5n}{3} + 7 = -12$ then $n =$

(A) –3 (B) $-8\frac{3}{5}$ (C) $-11\frac{2}{5}$ (D) $-31\frac{2}{3}$ [1]

12 In this quadrilateral $\angle KLM = 68°$. What is the size of $\angle NKM$?

(A) 34° (B) 44° (C) 56° (D) 68° [1]

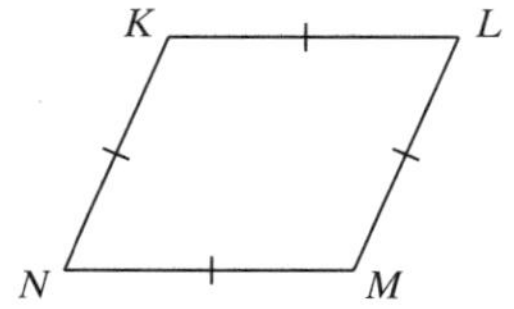

Marks

13 Which is **not** a Pythagorean triad?

Ⓐ 69, 260, 269 Ⓑ 105, 208, 233 Ⓒ 125, 238, 307 Ⓓ 204, 253, 325 — 1

14 Which is **not** a correct factorisation of $20p - 24pq$?

Ⓐ $-4p(5 - 6q)$ Ⓑ $2p(10 - 12q)$ Ⓒ $-4(6pq - 5p)$ Ⓓ $p(20 - 24q)$ — 1

15 If \$56 increases to \$182, what is the percentage increase?

Ⓐ 131% Ⓑ 225% Ⓒ 231% Ⓓ 325% — 1

16 The numbers from 1 to 20 are written on 20 cards, one number on each card. If one card is chosen at random, what is the probability that it is a multiple of 2 or 3 or both?

Ⓐ $\frac{13}{20}$ Ⓑ $\frac{7}{10}$ Ⓒ $\frac{3}{5}$ Ⓓ $\frac{1}{4}$ — 1

17 How many hectares are in 75 km^2?

Ⓐ 750 Ⓑ 7500 Ⓒ 75 000 Ⓓ 750 000 — 1

18 A kite has area 896 cm^2. One diagonal is 28 cm long. How long is the other diagonal?

Ⓐ 128 cm Ⓑ 64 cm Ⓒ 32 cm Ⓓ 16 cm — 1

19 What is the gradient (slope) of the line $y = 3 - x$?

Ⓐ -1 Ⓑ 1 Ⓒ 3 Ⓓ 0 — 1

20 Which triangle must be congruent to ΔPQR?

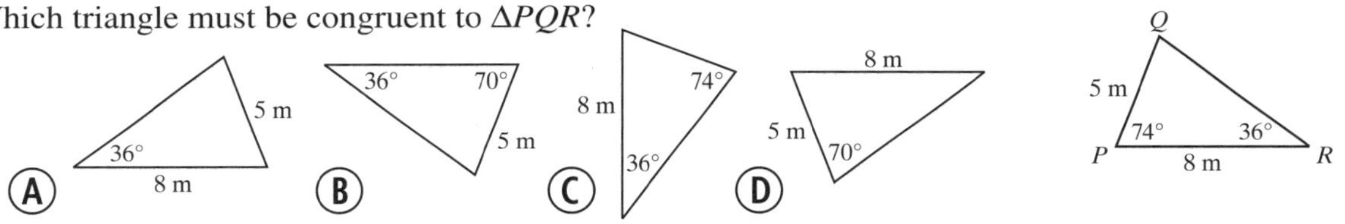

1

21 $8^{12} \div 2^6 =$

Ⓐ 4^6 Ⓑ 4^2 Ⓒ 2^6 Ⓓ 2^{30} — 1

22 $0.\dot{3}6\dot{3} =$

Ⓐ $\frac{4}{11}$ Ⓑ $\frac{11}{30}$ Ⓒ $\frac{37}{101}$ Ⓓ $\frac{121}{333}$ — 1

23 $\dfrac{9a^2 - 12a^2}{6a \times -a}$

Ⓐ $\frac{1}{2}$ Ⓑ $\frac{1}{2a^2}$ Ⓒ $11a^2$ Ⓓ 11 — 1

24 Which is equivalent to $3\frac{3}{5} : \frac{9}{10}$?

Ⓐ 5 : 1 Ⓑ 2 : 1 Ⓒ 11 : 3 Ⓓ 4 : 1 — 1

25 $PQRS$ is a parallelogram. What is the value of x?

Ⓐ 28 Ⓑ 30 Ⓒ 32 Ⓓ 34 — 1

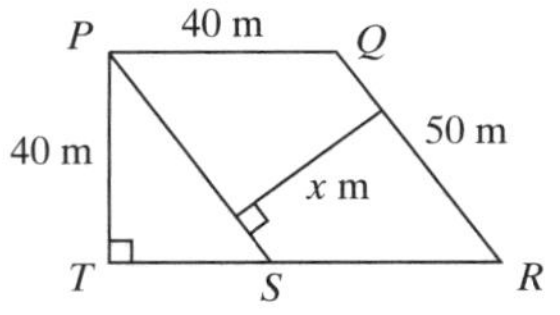

Total marks /25

EXAM PAPER 6 : LEVEL 3 — PART B

Marks

1 What is the value of 7π to one decimal place? __________ 1

2 What is $(3^4)^5 \div (3^0)^3$ in simplest index form? __________ 1

3 Find the percentage increase if \$15 is increased by \$20. __________ 1

4 $(12m^2n \times 7n) \div (21n^2 \times 2m) =$ __________ 1

5 What is the length of each side of a cube of capacity 729 mL? __________ 1

6 Malikah recorded the eye colour of all the students in her class. What type of data is this? __________ 1

7 Evaluate $\frac{(-6)^3 - (-3)^4}{-72 + 18}$. __________ 1

8 The ratio of red to blue balloons is 5 : 3. If there are 96 red and blue balloons altogether, how many more red than blue balloons are there? __________ 1

9 The probability of winning a game is $\frac{1}{5}$ and the probability of a draw, tie or no result is $\frac{2}{7}$. What is the probability of losing the game? __________ 1

10 If $a = -7$, $b = -8$ and $c = -9$, what is the value of $c^2 - ab$? __________ 1

11 A rhombus of side length 15 cm has diagonals of length 24 cm and 18 cm. What is its area? __________ 1

12 $-5x^2 - 3x + 7x^2 - 12x - 9x^2 =$ __________ 1

13 A movie that runs for 2 hours 47 minutes begins at 11:35 am. When does it finish? __________ 1

14 Express $\frac{4}{13}$ as a recurring decimal. __________ 1

15 Factorise fully: $9m^2n - 6mn^2$. __________ 1

16 If there is one chance in a million of being born with a certain condition, what is the probability of **not** being born with that condition? __________ 1

17 If one angle of a parallelogram measures 58°, what are the sizes of the other angles? __________ 1

Marks

18 $(43 - 11 \times -5) \div (-17 - -38) =$ ______ 1

19 What is the highest score of a data set if the lowest score is 37 and the range is 56? ______ 1

20 What is the solution to $9x - 7 = 4x + 7$? ______ 1

21 What is the value of x? (Leave the answer as a square root.) ______ 1

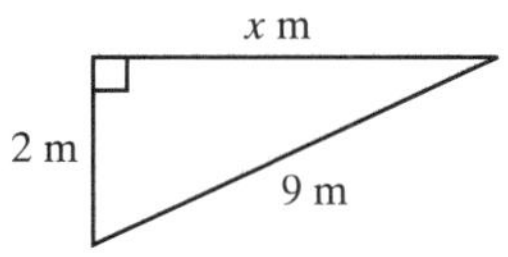

22 A ring was sold for $1600 which was a profit of 25%. What was the cost price? ______ 1

23 $-3x(5x - 2y + z) =$ ______ 1

24 A map has a scale of 1 : 100 000. What is the actual distance between two places that are 3.6 cm apart on the map? ______ 1

25 Write an equation for the points in the table of values. ______ 1

x	1	2	3	4	5
y	4	16	28	40	52

26 At 75 km/h, how long, in hours and minutes, will it take to travel 315 km? ______ 1

27 What fraction is $0.\dot{2}\dot{4}$? ______ 1

28 What is the radius, to one decimal place, of a circle of circumference 53.4 cm? ______ 1

29 How many square centimetres is 144 m^2? ______ 1

30 What is the volume of a prism of perpendicular height 37 cm, if the cross-section is a right-angled triangle with sides 32 cm, 60 cm and 68 cm? ______ 1

Total marks /30

1 Expand and simplify $5p(3p - 4) - 6(p^2 + 2) - 8(7 - p)$. 2 marks

2 Some members of a representative team received gold, silver and bronze medals. The numbers who received at least one of each type of medal are shown in the Venn diagram.

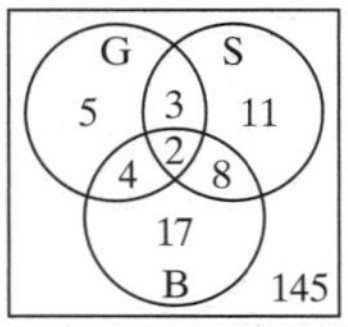

a How many members of the team are there? 1 mark

b How many members received at least one gold medal? 1 mark

c What is the probability that a member of the team, chosen at random, had medals of exactly two different types? 1 mark

d What is the probability that a medal winner had a silver or bronze medal, but not gold? 1 mark

3 **a** What is the percentage increase if \$3200 is increased by \$24? 1 mark

b If \$960 is increased by the same percentage, what is the new amount? 1 mark

c What amount would increase to \$2035.15 at the same percentage? 1 mark

4 The diagram shows the graphs of $y = x^2$ and a straight line.

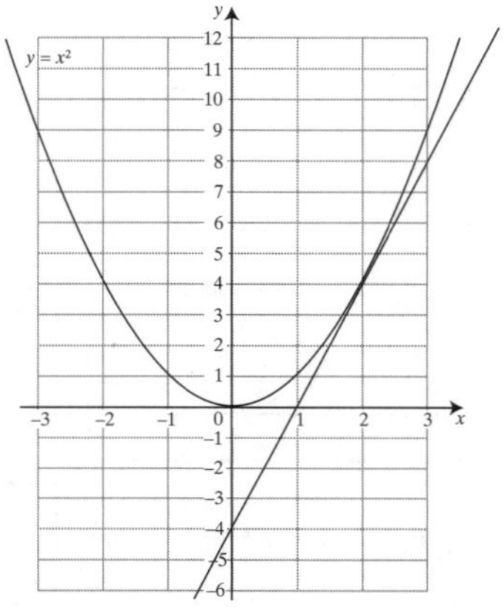

a Briefly explain why $y = x^2$ is not a linear relationship. 1 mark

b At what point does the line cross the x-axis? 1 mark

c What is the y-intercept of the line? 1 mark

d What is the gradient of the line? 1 mark

e What is the equation of the line? 1 mark

f What is the equation of the line that passes through the origin and is parallel to the first line? 1 mark

5 **a** Find $5\frac{2}{3} - 2\frac{3}{4} \div -1\frac{1}{5}$ giving the answer as a mixed number in simplest form. 1 mark

b Express the answer to part **a** as a decimal. 1 mark

6 An athletics field is made up of a rectangle and two semicircles, as shown in the diagram. A running track goes around the outside of the field.

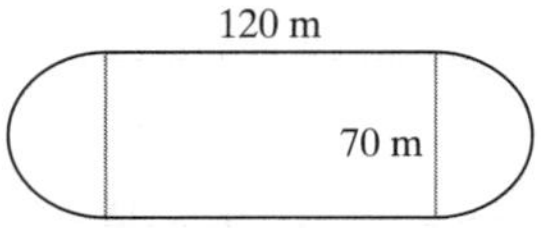

a How far is it around the track, to the nearest metre? 2 marks

b How long, in minutes and seconds, would it take to run five laps of the track at a speed of 4 metres per second? 2 marks

7 'When thirteen is subtracted from a certain number and that result is multiplied by eight the answer is eleven more than three times the number.'

a Let the number be x. Write an equation from the given information. 1 mark

b Find the number. [2 marks]

8 $ABCD$ is a kite. $AB = AD$ and $BC = DC$.

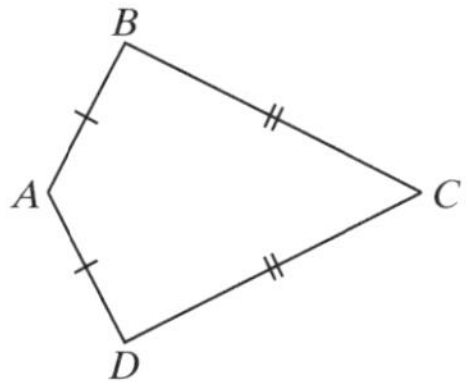

a Describe the transformation that would move ΔABC onto ΔADC. [1 mark]

b Which test shows that $\Delta ABC \equiv \Delta ADC$? [1 mark]

c Explain why diagonal AC must bisect the angles of the kite that it passes through. [1 mark]

d What other result about the angles of a kite follows from the fact that the triangles are congruent? [1 mark]

e If BD also bisected the angles of the kite through which it passes, what would that mean about $ABCD$? [1 mark]

9 A car leaves home at 8:45 am and returns at 1:20 pm.

a How long, in hours and minutes, did the trip take? [1 mark]

b During this time, the car travelled a distance of 340 km and was stopped for 20 minutes. What was the average speed when the car was moving? [1 mark]

c If the car used petrol at the rate of 7.5L/100 km, how much petrol did it use on the trip? [1 mark]

10 Find the capacity, in litres, of this container. [2 marks]

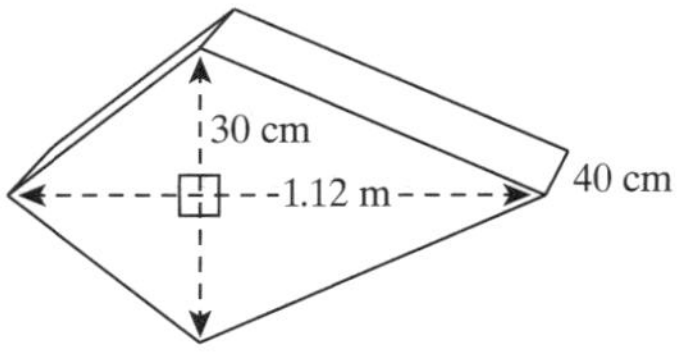

11 The graph shows the minimum daily temperature for a town for July.

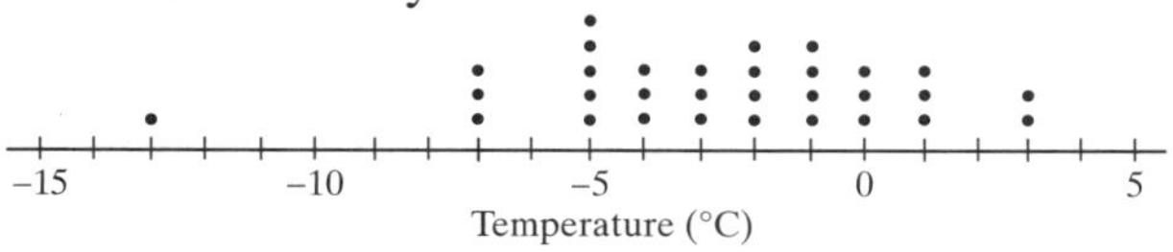

a What is the most common temperature? [1 mark]

b What is the median temperature? [1 mark]

c Discuss the spread of the temperatures, mentioning any outliers. [2 marks]

d What is the mean temperature if the lowest temperature is disregarded? [1 mark]

12 The diagram shows a square in a circle of radius 12 cm.

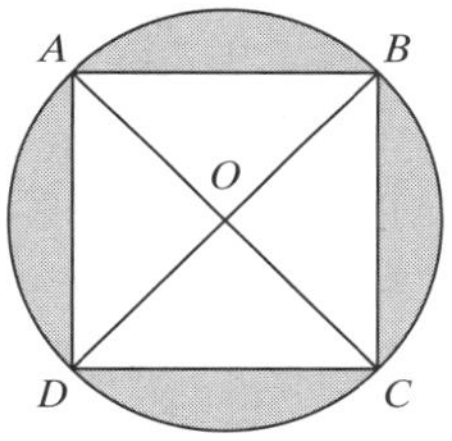

a What is the length of each diagonal of the square? [1 mark]

b What is the area of the square? [1 mark]

c Explain why ΔAOB is both isosceles and right-angled. [1 mark]

d Find the length of AB. Give the answer to one decimal place. [1 mark]

e What is the area of the circle in terms of π. [1 mark]

f Find the shaded area to the nearest square centimetre. [1 mark]

Total marks for PART C /45 **Total marks** /100

Answers

The Four Operations with Integers

Page 1 1 D **2** D **3** C **4** A **5** B **6** C **7** D **8** C **9** B **10** B **11** A **12** A **13** D **14** C **15** B **16** A

Page 2 **1** D **2** A **3** B **4** D **5** B **6** A **7** A **8** D **9** B **10** C **11** C **12** A **13** B **14** C **15** D

Page 3 **16** –10 **17** –5 **18** –18 **19** –23 **20** –5 **21** –12 **22** –11 **23** –7 **24** 70 **25** –8 **26 a** 1776 **b** –48 **c** –37 **27 a** 2 **b** –44 **c** –14 **28 a** 2 °C **b** 7 °C **c** They are all the same. **d** –14 °C **e** 13 °C

Page 4 **1** A **2** A **3** D **4** C **5** D **6** C **7** A **8** D **9** C **10** A **11** B **12** D **13** B **14** B **15** B

Page 5 **16** 42 **17** –38 **18** 360 **19** –32 **20** 14 **21** –18 **22** 12 **23** –6 **24** –2 **25** –26 **26 a** –3024 **b** 3024 **c** –6 **d** –72 **27 a** 5 °C **b** 8 hours **28 a** –1535 **b** 3 **c** –50 **d** –10 700

Page 6 **1** A **2** B **3** B **4** A **5** D **6** C **7** A **8** C **9** D **10** B **11** D **12** D **13** B **14** C **15** C

Page 7 **16** –44 **17** –164 **18** 29 **19** –5 **20** –65 **21** –13 **22** –276 480 **23** 7 **24** 9 **25** –6 **26 a** The business had made a loss. **b** \$773 **c** \$3808 **d** –\$1308.75 **e** \$5235 **27 a** 26 **b** –32 **c** $-7 + (9 \times 5) - (-6 \times -2)$ **28 a** $-48 \div (6 \times -2) = 4$ **b** $(7 - -5) \times (4 - 6 - 3) = -60$ **c** $-3 \times -8 - (6 - -1 + 12 \div -4) = 20$ **29 a** $-\frac{1}{3}$ **b** Error message; the denominator is zero and it is not possible to divide by zero.

Rational and Irrational Numbers

Page 8 **1** D **2** C **3** B **4** B **5** C **6** C **7** D **8** B **9** A **10** A **11** D **12** A **13** B **14** C **15** D

Page 9 **1** D **2** C **3** D **4** A **5** B **6** A **7** B **8** A **9** C **10** D **11** D **12** B **13** C **14** A **15** B

Page 10 **16** irrational **17** 0.3556 **18** terminate **19** $\frac{2}{3}$ **20** rational **21** –30 **22** rational **23** $0.1\dot{6}$ **24** irrational **25** $-\frac{4}{5}$ **26 a** 0.096 969 696 9696… **b** 0.0970 **27 a** $3\frac{3}{4}$ **b** 3.75 **c** It can be expressed as a fraction. **28 a** 2.6 **b** $2\frac{3}{5}$ **c** $\frac{13}{5}$ **d** $\frac{1}{5}$, 0.2 **29** It is not correct. $\sqrt{3}$ is irrational so cannot be expressed as a recurring decimal. [$\sqrt{3} = 1.732\,050\,807\,568\ldots$] **30** $0.\dot{6}\dot{3}$

Page 11 **1** D **2** C **3** A **4** A **5** B **6** D **7** C **8** C **9** A **10** C **11** B **12** D **13** D **14** B **15** A

Page 12 **16** $\frac{4}{11}$ **17** 0.625 **18** repeat **19** 3.605 55 **20** irrational **21** 91.8 **22** 0.047 475 **23** –5.65 **24** $1.\dot{4}$ **25** $-7\frac{7}{24}$ **26 a** $\frac{58}{111}$ **b** $0.\dot{5}2\dot{2}$ **c** 0.522 523 **27 a** $-0.0\dot{6}$ **b** $-\frac{1}{15}$ **28 a** Yes, any number that can be expressed as a fraction is rational. **b** 0, the number has been rounded up **c** 16, there are 16 different possible remainders when dividing by 17. Once the same remainder occurs, the answer will begin to repeat. **29 a** $-5\frac{127}{360}$ **b** $-5.352\dot{7}$ **c** –5.3528 **d** $0.0000\dot{2}$

Indices

Page 13 **1** C **2** B **3** B **4** A **5** B **6** C **7** D **8** A **9** B **10** C **11** D **12** A **13** C **14** D **15** D **16** A

Page 14 **1** A **2** D **3** B **4** B **5** C **6** C **7** D **8** B **9** D **10** A **11** C **12** C **13** D **14** B **15** A

Page 15 **16** 2^{15} **17** 3^{10} **18** 2^{20} **19** 1 **20** –8 **21** 5^9 **22** 10^5 **23** 3^{13} **24** 5 **25** 7^{10} **26 a** 3^2 **b** 5^{28} **c** 2^{41} **27** 5 **28 a** 3^4 **b** 5^6 **c** 2^{12} **29 a** 8 **b** 1 **c** Using the index laws we subtract the indices and get 2^0. But, from part **b** $2^3 \div 2^3 = 1$. So as the answers are to the same question it follows that $2^0 = 1$.

Page 16 **1** A **2** A **3** C **4** B **5** C **6** A **7** D **8** A **9** B **10** D **11** C **12** B **13** C **14** B **15** D

Page 17 **16** 3^{10} **17** 33 **18** 1 **19** 11^6 **20** 3^{35} **21** 7^{11} **22** 6^3 **23** 5^7 **24** 10^7 **25** 2^{12} **26 a** 2^{32} **b** 3^{10} **c** 2^{12} **d** 5^7 **e** 5^3 **27 a** 42 **b** 8 **c** 10 000 000

Page 18 **1** D **2** C **3** B **4** D **5** A **6** A **7** C **8** D **9** A **10** C **11** B **12** B **13** A **14** C **15** D

Page 19 **16** 13 **17** 2^{16} **18** 2^{14} **19** 1 **20** 7 **21** –27 **22** 6^3 **23** 32 **24** 5^{14} **25** 17^{11} **26 a** 3^8 **b** 2^{16} **c** 7^{18} **27 a** 81 **b** 1 **c** 7 200 000 **28 a** $2^2 \times 3^3$ **b** $2^8 \times 3^{12}$ **29** Nancy multiplied the base numbers and multiplied the indices. She should not have changed the base number and added the indices. The correct answer is 3^8. There are 5 factors of 3 multiplied by 3 factors of 3 so altogether there are 8 factors of 3.

Percentages

Page 20 **1** C **2** A **3** B **4** D **5** D **6** C **7** A **8** C **9** B **10** A **11** D **12** B **13** A **14** C **15** D

Page 21 **1** D **2** B **3** A **4** C **5** B **6** D **7** A **8** B **9** D **10** A **11** D **12** C **13** C **14** C **15** B

Page 22 **16** \$280 **17** 40% **18** \$800 **19** \$144 **20** \$2700 **21** \$240 **22** 20% **23** \$2160 **24** 10% **25** 10% **26 a** \$300 **b** 25% **c** \$1000 **27 a** \$300 000 **b** 150% **28 a** 85% **b** 30 **29 a** 15% **b** \$625 **c** \$100 **30 a** \$525 **b** 8%

Page 23 **1** C **2** D **3** D **4** A **5** B **6** C **7** D **8** A **9** B **10** C **11** C **12** A **13** B **14** A **15** B

Page 24 **16** \$13.68 **17** 72.5% **18** \$600 **19** 24 m **20** \$819 **21** \$1890 **22** 20% **23** \$24 000 **24** \$4800 **25** 25% **26** 11.8% **27 a** \$2960 **b** \$24 420 **28** 21 **29 a** \$144 **b** \$2016 **c** \$2100 **d** 30% **e** 32%

Page 25 1 A 2 D 3 B 4 B 5 B 6 C 7 C 8 D 9 A 10 B 11 A 12 C 13 A 14 D 15 C

Page 26 **16** $33\frac{1}{3}$% **17** \$162 000 **18** $58\frac{1}{3}$% **19** 61.5 ha **20** 210 **21** 7.5% **22** 87.75 **23** \$75 **24** $133\frac{1}{3}$% **25** \$176 **26 a** 12% **b** \$2816 **c** 140% **d** \$65 **27 a** 52% **b** 2700 **c** 79 412, 38 118 **28 a** 720 000 **b** 18% **29 a** \$2760 **b** The loss is 15% of a higher amount (\$2760 not \$2400) so Annie's loss will be greater than the profit Ed made. **c** \$54 less

Ratio and Rates

Page 27 1 B 2 D 3 C 4 A 5 B 6 C 7 D 8 C 9 C 10 D 11 A 12 D 13 B 14 A 15 B

Page 28 1 B 2 A 3 C 4 D 5 D 6 B 7 C 8 B 9 A 10 A 11 C 12 C 13 B 14 A 15 D

Page 29 **16** \$50 **17** 102 km **18** 1250 km **19** 2 : 3 : 1 **20** 180 **21** 1 h 15 min **22** 3 : 2 **23** 12.5 m **24** 12 km/h **25** 49 **26 a** 9 **b** 1 : 2 **27 a** 60 km/h **b** \$15 **c** 2 h **28 a** 4 : 1 **b** 16 : 3 : 1 **29 a** \$1080 **b** \$80 **c** 50

Page 30 1 B 2 A 3 B 4 B 5 C 6 D 7 B 8 D 9 C 10 C 11 A 12 C 13 D 14 D 15 A

Page 31 **16** 7 : 5 : 8 **17** 65 km/h **18** \$7200 **19** 3 h 15 min **20** 252 **21** 15 : 2 **22** 4.8 t **23** 1560 **24** 3 : 5 **25** 264 km **26** Ratios compare quantities of the same kind while rates compare quantities of different kinds. **27 a** 1 : 400 **b** 8.5 cm **c** 54 m **28 a** 522 km **b** 72 km/h **c** 39.15 L **d** 7.5 L/100 km **29 a** 3.75 L/min **b** 337.5 L **c** 4 d 10 h 40 min **30** 24 kg rye, 7.2 kg chicory, 4.8 kg clover

Page 32 1 B 2 A 3 C 4 C 5 D 6 B 7 A 8 D 9 B 10 C 11 D 12 A 13 C 14 B 15 D

Page 33 **16** 18 : 1 **17** 3 h 22.5 min **18** 10 : 1 **19** 198 **20** 5 : 24 **21** 160 g **22** \$380 **23** \$1110 **24** 54 **25** 1 : 2 **26 a** 3.5 km **b** 17.5 cm **c** 5 : 1 **d** 210 km/h **27 a** 6 **b** 10 **28 a** 17.46 km/h **b** 2 h 54 min **c** about 7 km **29 a** 2.25 m^3 **b** 0.75 m^3 sand, 1.125 m^3 aggregate **c** 0.003 125 m^3

Mixed Number

Page 34 1 A 2 B 3 B 4 C 5 A 6 C 7 B 8 C 9 A 10 C 11 D 12 D 13 B 14 A 15 D

Page 35 **16** 0.667 **17** \$360 **18** 22 **19** –4 **20** 28 **21** 12 : 6 : 1 **22** $\frac{1}{3}$ **23** 60 km **24** 150% **25** 2^{24} **26 a** –40 736 **b** 2144 **c** –32 **27** $0.58\dot{3}$ **28 a** \$32 **b** 15% **c** \$2 **29 a** 80 **b** No, the ratio will not change because the numbers sold are in the same ratio as the original number. **30 a** 70 km/h **b** 33.6 L **31** No, the answer should be 7^7. The base numbers should not have been multiplied. Three factors of 7 multiplied by 4 factors of 7 is 7 factors of 7.

Page 36 1 B 2 A 3 C 4 B 5 A 6 D 7 D 8 B 9 C 10 B 11 A 12 D 13 C 14 A 15 C

Page 37 **16** 3.1416 **17** $\frac{25}{33}$ **18** 150% **19** 20 : 15 : 12 **20** 30 **21** 30% **22** 3 h 24 min **23** 7^{11} **24** 0.06 **25** 1320 g **26 a** 20 L/min **b** 1 h 37 min **c** $1.61\dot{6}$h **27 a** 12% **b** \$3864 **c** \$1600 **28** 22 : 21 **29 a** 22 °C **b** –3 °C **30 a** \$402.50 **b** 20 : 23 **c** \$53.75 **c** 20 : 43

The Four Operations with Algebra

Page 38 1 B 2 A 3 B 4 D 5 B 6 D 7 C 8 D 9 D 10 B 11 A 12 C 13 C 14 A 15 A

Page 39 1 B 2 B 3 D 4 C 5 A 6 D 7 B 8 D 9 A 10 D 11 A 12 C 13 C 14 A 15 B

Page 40 **16** $6x$ **17** b **18** $20x^2$ **19** $9 - 6m$ **20** $11k^2$ **21** $11p$ **22** $5a$ **23** $2x + 8y$ **24** $-80ab$ **25** $90x^2$ **26 a** 48 **b** 44 **c** 10 **27 a** x^2 **b** $a + 11b + 1$ **c** $20m$ **d** $14n + 3n^2$ **e** 14 **f** $23p^2 - 4p$

Page 41 1 B 2 C 3 A 4 C 5 A 6 B 7 A 8 B 9 D 10 D 11 A 12 B 13 C 14 C 15 D

Page 42 **16** $4x^2$ **17** $11m^2$ **18** $3xz^2$ **19** $12m - 3n$ **20** $4a^2 + 4a$ **21** $-12a^2bc$ **22** $36d^2$ **23** 11 **24** $\frac{b}{3a}$ **25** $10a^2bc$ **26** 12 **27 a** $48x^2$ **b** $8b^2c + 10ab$ **c** $5m - 3n - 16$ **d** 2 **e** –2 **28 a** –9 **b** 16 **c** 1

Page 43 1 C 2 A 3 B 4 C 5 B 6 C 7 A 8 D 9 D 10 D 11 A 12 B 13 D 14 A 15 B

Page 44 **16** $2p^2$ **17** $-4x^2 + 13x$ **18** $-9a$ **19** $28xy - 6x + y$ **20** 12 **21** $55a^2$ **22** 5 **23** $6x^2$ **24** $13t + t^2$ **25** $-3 - 8d$ **26 a** $2x + 3y$ **b** $20ab$ **c** $100x$ **27** –6 **28 a** –2 **b** $-14a^2b$ **c** $2x$ **d** $-88p^2r^2$ **29** 225; Nell should have put $(-15)^2$ but instead the calculator worked out $-(15^2)$.

Expanding and Factorising

Page 45 1 C 2 A 3 B 4 C 5 B 6 D 7 B 8 A 9 C 10 D 11 D 12 A 13 B 14 C 15 D 16 A

Page 46 1 B 2 C 3 B 4 C 5 A 6 A 7 D 8 C 9 A 10 D 11 C 12 D 13 A 14 B 15 D

Page 47 **16** $5a + 30$ **17** $2 \times 5 \times 7$ **18** 6 **19** $14x - 7$ **20** $15a + 3b$ **21** $3(x + 10)$ **22** $2(7a - 6)$ **23** $a^2 + 4a$ **24** $x(x - 9)$ **25** $36m - 24n$ **26** $-12a - 8b - 15c + 10d$ **27** They are both correct, but neither has fully factorised the expression. Fully factorised $24x + 36 = 12(2x + 3)$. **28 a** $35x + 18y$ **b** $20n - 31$ **c** $x^2 + 2xy + y^2$ **d** $5a + 25$ **29 a** $3x(3x - 4y)$ **b** $8a(2bc + 1)$

Page 48 **1** C **2** B **3** D **4** D **5** B **6** D **7** B **8** A **9** B **10** C **11** C **12** D **13** C **14** A **15** A

Page 49 **16** $45x - 36$ **17** $a^2 - 7a$ **18** $20x + 30y - 40z$ **19** $2(8a + 9b)$ **20** $-8x - 6$ **21** $-15m + 6n$ **22** $4x(3x - 5y)$ **23** $x - 7$ **24** $-x + y$ **25** $14a(2b^2 + a)$ **26 a** $-6n - 5$ **b** $9a^2 - 2a$ **c** $-5u$ **d** $6x^2 + 2x - 14$ **27** Both are correct because both answers when expanded give $9b - 3a$. **28** 260 **29 a** $7xy(6x - 4y + 5xy)$ **b** $5a^2b^3c^2(2a - 5c)$

Page 50 **1** B **2** B **3** C **4** A **5** B **6** A **7** D **8** D **9** C **10** B **11** A **12** C **13** D **14** A **15** C

Page 51 **16** $3xyz$ **17** $5k^2 - 5kl + 35k$ **18** $9x(1 - 5y)$ **19** $-8a^2 + 14a$ **20** $-11 - k$ **21** $ab^2 + b^2c$ **22** $-3p^2 - 4p$ **23** $4y^2(2x + 1)$ **24** $a^2b - ab^2$ **25** -2 **26 a** $2x^2 + 9x$ **b** $x(2x + 9)$ **c** x is a common factor. $x(x + 4 + x + 5) = x(2x + 9)$ **27 a** $27n - 52$ **b** $8m^2 + 8m - 11$ **c** 76 **28 a** $3pq^2(4p^2 - 3pq - 5)$ **b** $a^2b^2c^2(abc + b + ac + 1)$ **c** $4x(x + 2y - 6y^2 + 1 - 5xy + 4x^2)$

Equations

Page 52 **1** A **2** C **3** C **4** C **5** A **6** A **7** D **8** D **9** B **10** A **11** B **12** C **13** B **14** D **15** B

Page 53 **1** A **2** C **3** B **4** C **5** D **6** B **7** A **8** D **9** B **10** C **11** B **12** A **13** D **14** D **15** A

Page 54 **16** $x = 4$ **17** false **18** 0.625 **19** $x = 24$ **20** true **21** $x = 10$ **22** $x = 6$ **23** 20 **24** 1 **25** $x = 0$ **26 a** $x = 2$ **b** $x = 5\frac{1}{3}$ **c** $x = 59$ **d** $x = -5$ **e** $x = 3$ **f** $x = -48$ **27 a** $x = 3$ or $x = -3$ **b** The square of any real number must always be positive, so no number squared will equal a negative number.

Page 55 **1** D **2** B **3** C **4** A **5** A **6** A **7** B **8** D **9** C **10** B **11** A **12** D **13** B **14** C **15** C

Page 56 **16** $x = 7$ **17** $x = -4$ **18** true **19** 6 **20** $x = -50$ **21** $x = -18$ **22** true **23** $x = 9$ **24** false **25** 77 **26 a** $x = 4$ **b** $x = 29$ **c** $x = -\frac{2}{3}$ **d** $x = 1.75$ **e** $x = 5\frac{1}{3}$ **f** $x = -0.9$ **27 a** $4(5n + 17) = 508$ **b** The number is 22.

Page 57 **1** C **2** B **3** A **4** A **5** D **6** D **7** D **8** C **9** A **10** B **11** B **12** C **13** B **14** C **15** A

Page 58 **16** $x = 7$ **17** $x = 9$ **18** 855 **19** $x = -4$ **20** -3.8 **21** 2.65 **22** $m = 22$ **23** $p = -6$ **24** true **25** $n = -340$ **26 a** $p = -1.3$ **b** $m = 2.4$ **c** $a = \frac{13}{22}$ **d** $n = -1\frac{7}{15}$ **e** $v = 13\frac{21}{23}$ **27** $x = 0.19791\dot{6}$ **28 a** $7C + 2.55 = 4C + 7.65$ **b** \$1.70 **c** \$14.45

Linear Relationships

Page 59 **1** B **2** C **3** D **4** B **5** C **6** A **7** C **8** C **9** D **10** D **11** B **12** D **13** A **14** A

Page 60 **1** A **2** C **3** C **4** B **5** C **6** B **7** C **8** D **9** B **10** B **11** D **12** D **13** D **14** A **15** A

Page 61 **16** 9 **17** 4 **18** -3 **19** 0 **20** 22 **21** $y = 7x + 8$ **22** 4 **23** (4, 14) **24** $x = 6$ **25** $x = 1$ **26 a** 16, 21, 26, 31 **b** one more than five times the number of hexagons **c** $m = 5h + 1$ **d** 251 **e** 60 **27 a** 3, 5, 7, 9 **b** (below) **c** $x = -2$ **28 a** All pass through the point (0, 1). **b** All have the same constant term, 1 (the y-intercept). **c** It determines where the line intercepts the y-axis. **d** The slope (gradient) is negative, so the line will decrease.

27 b

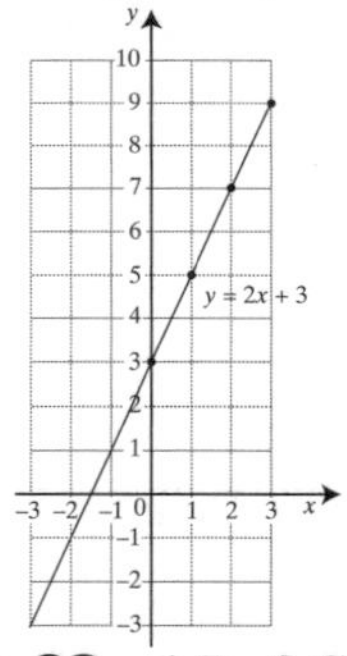

Page 63 **1** B **2** C **3** D **4** B **5** D **6** C **7** C **8** C **9** D **10** A **11** D **12** A **13** B **14** A **15** A

Page 64 **16** 91 **17** 6 **18** $y = 6x + 2$ **19** (0, 5) **20** $y = 7x + 2$ **21** increase **22** false **23** 20 **24** false **25** 6 **26 a** 26, 33, 40, 47 **b** The number of matches is two less than seven times the stage number. **c** $m = 7s - 2$ **d** 103 **e** 26 **27 a** $-7, -5, -3$ **b** 5, 4, 3 **c** (below) **d** (4, 1) **e** $x = 6$ **28 a** All have the same coefficient of x. The gradient of each one is 4. The lines are parallel. **b** It gives the value of the y-intercept, where the line cuts the y-axis.

27 c

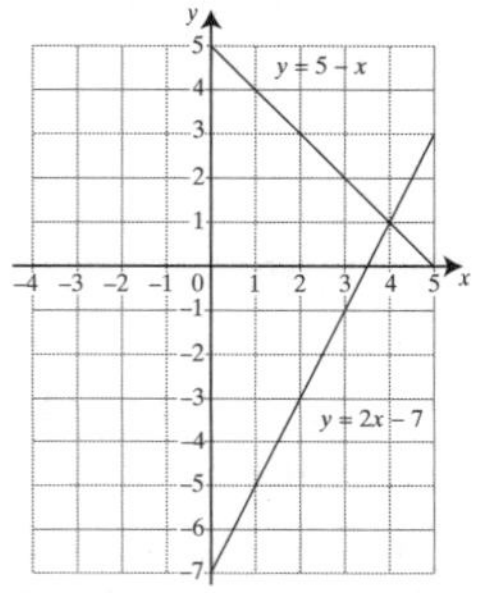

Page 65 **1** A **2** B **3** C **4** D **5** B **6** A **7** C **8** C **9** A **10** C **11** A **12** B **13** D **14** B **15** D

Page 66 **16** 10 **17** true **18** 25 **19** (0, –2) **20** 17 **21** –3 **22** false **23** 5 **24** –9 **25** $y = 8x - 5$ **26 a** 24, 28 **b** $p = 4s + 8$ **c** 56 **d** Stage 24 **27 a** –3, 1, 5 **b** (below) **c** an infinite number **d** No, they are parallel. **28 a** 3 **b** It is a measure of the steepness of the line. From any point, for every 1 unit moved right, moving up 3 units will give another point on the line. **c** The line decreases. Moving right from a point and down will find another point. **29 a** No, it is not a straight line. **b** 3 **c** (–1, 1) and (3, 9) **d** $y = 2x + 3$

27 b

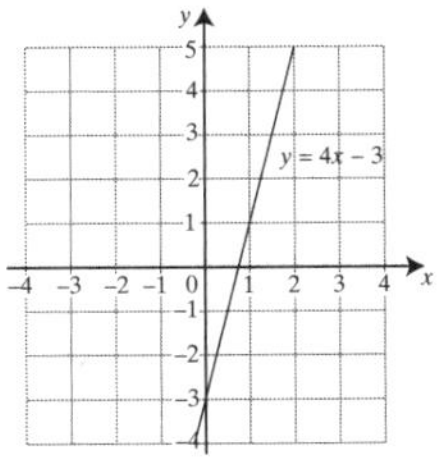

Mixed Algebra and Equations

Page 68 **1** A **2** A **3** B **4** C **5** D **6** B **7** D **8** D **9** B **10** B **11** C **12** A **13** D **14** C **15** A

Page 69 **16** $15x - 6$ **17** $11x + 4y$ **18** $3(x - 5)$ **19** $14a$ **20** $x = 14$ **21** $x^2 + xy$ **22** $a(a + 2)$ **23** 12 **24** $10q$ **25** $-8x - 10$ **26 a** $2x + 12x^2$ **b** $2a$ **27 a** $x^2 + 6x + 6$ **b** $23a - 6$ **28 a** $5x(x - 2)$ **b** $4(2a + 1)$ **29 a** $x = 6$ **b** $x = 5$ **30 a** 8, 6, 4, 2 **b** (right) **c** $x = 5$

30 b

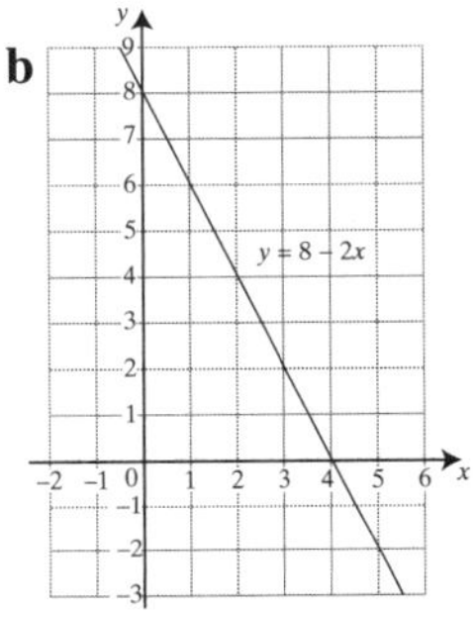

Page 70 **1** A **2** B **3** A **4** D **5** B **6** D **7** C **8** A **9** C **10** D **11** B **12** D **13** A **14** B **15** C

Page 71 **16** $x = 20$ **17** $2x^2 - 8x$ **18** $4(2a - 1)$ **19** 1 **20** 3 **21** $4x^2 + 2x$ **22** 977 **23** $12m^2$ **24** $-3a$ **25** $a + 2$ **26 a** $x = -10$ **b** $x = 16$ **27 a** $2x^2 + x - 7$ **b** $p^2 + 27p$ **28** $4ab(2a + 3b + 5)$ **29 a** 3, 4, 5, 6 **b** –1, –2, –3, –4 **c** (below) **d** $x = -2$

29 c

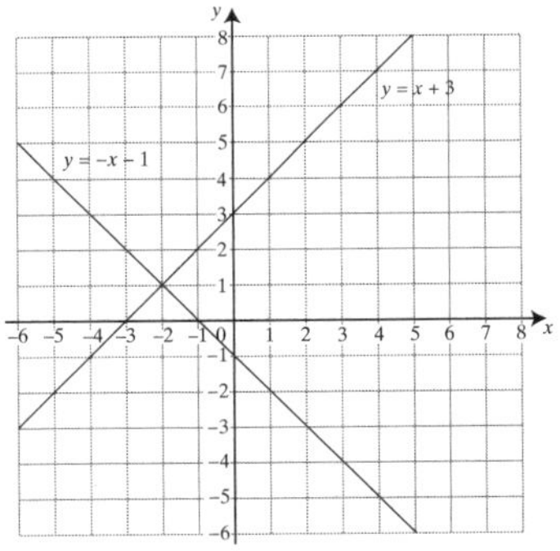

Page 72 **1** A **2** D **3** D **4** C **5** A **6** B **7** A **8** B **9** D **10** D **11** C **12** B **13** A **14** B **15** C

Page 73 **16** $3(a^2 + 2a + 1)$ **17** $-p^2 + 2p - 2q$ **18** $-6a^2 + 10abc$ **19** $p = 5$ **20** $y = 7x - 11$ **21** 1 **22** $a^2b^2c - ab^2c^2$ **23** (0, –3) **24** –13 **25** $-3m(2n + 1)$ **26 a** $22p^2q + 15pq^2 - 20p$ **b** $8x^2 - 46xy + 21y^2 - 2xy^2$ **27 a** $x = -3\frac{1}{4}$ **b** $x = -11\frac{3}{7}$ **28** Neither fully factorised the expression. Both answers could be factorised further. $12xyz(2z + 3x)$ **29 a** (4, 0) **b** It is the only point that satisfies both equations. **c** (right) **d** It is parallel to the line $y = 2x - 8$ and has the same coefficient of x and the same gradient, 2. It has the same y-intercept as $y = 4 - x$ and the same constant term.

29 c

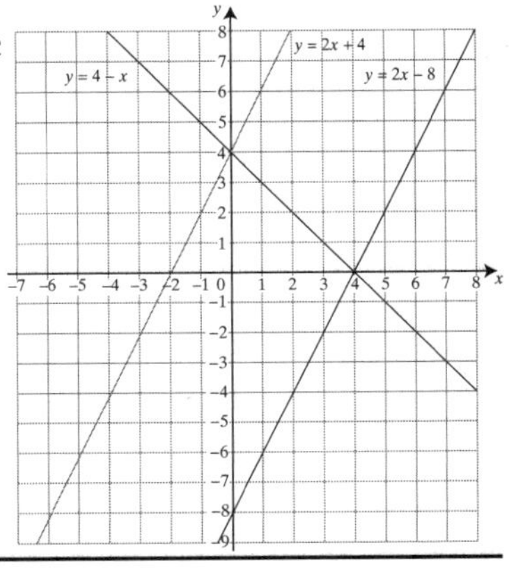

Perimeter and Area

Page 74 **1** D **2** B **3** C **4** B **5** D **6** C **7** A **8** C **9** A **10** A **11** C **12** B **13** B **14** D **15** D **16** A
Page 75 **1** A **2** B **3** C **4** D **5** B **6** D **7** A **8** C **9** C **10** B **11** D **12** C **13** D **14** B **15** A
Page 76 **16** 40 000 **17** 96 cm^2 **18** 31.5 cm^2 **19** 9 000 000 **20** 78 m **21** 3.86 m **22** 8680 cm^2 **23** 90 cm^2 **24** 4 m **25** 0.768 m^2 **26 a** \$6864 **b** \$117 **27 a** 8148 cm^2 **b** 1.97 m **28 a** Find the area of one of the trapeziums that make up the frame and multiply by 4; find the area of the inner square and subtract from the area of the outer square. **b** 576 cm^2 **29 a** 34 cm **b** 60 cm^2 **c** 7.1
Page 77 **1** B **2** B **3** B **4** C **5** D **6** C **7** D **8** D **9** C **10** A **11** A **12** D **13** A **14** A **15** B
Page 78 **16** 200 000 **17** 256 cm^2 **18** 3.12 m **19** 0.4876 m^2 **20** 313.2 cm^2 **21** 86 cm **22** 65 cm **23** 75 cm **24** 900 **25** 288 cm^2 **26 a** 7675 m^2 **b** 350 **27 a** 146 m **b** 220 m **c** 192 m **d** 21 120 m^2 **28** 102 cm^2 **29 a** 135.36 cm^2 **b** 13.64
Page 79 **1** C **2** B **3** A **4** D **5** C **6** D **7** B **8** A **9** D **10** C **11** A **12** B **13** A **14** C **15** B
Page 80 **16** 37.5 **17** 49 cm **18** 10 cm **19** 50 **20** 601 m **21** 53.7 ha **22** 361 cm^2 **23** 7 cm **24** 234 cm^2 **25** 16 cm **26 a** 3.5178 m^2 **b** 30.16 m **27** 585 cm^2 **28** 3412.5 m^2 **29 a** 0.0099 m^2 **b** 7.3 cm **30 a** Multiplying the base by the height, using the base as 13.75 m and the perpendicular height of 25.2 m. **b** 9.3 **c** 41.25

Circles

Page 81 **1** B **2** C **3** C **4** D **5** B **6** B **7** B **8** C **9** D **10** C **11** A **12** D **13** A **14** A **15** D
Page 82 **1** B **2** D **3** B **4** C **5** D **6** D **7** C **8** A **9** A **10** B **11** C **12** A **13** A **14** B **15** C
Page 83 **16** 19.4 cm **17** 3.141 59 **18** 8π cm **19** 9π cm^2 **20** 9.4 m **21** 2124 cm^2 **22** 295 cm **23** 4071.5 km^2 **24** 55 m **25** 241 m^2 **26 a** $\frac{1}{6}$ **b** 151 cm^2 **c** 52 cm **27** 1896.7 cm^2 **28 a** 5 **b** 73 m **29 a** The length of the rectangle is equal to the diameter of the circle. The circle has radius of (12 – 5) m or 7 m. **b** 147 m^2 **c** 46 m
Page 84 **1** C **2** B **3** A **4** C **5** A **6** B **7** A **8** B **9** C **10** B **11** A **12** D **13** B **14** D **15** D
Page 85 **16** 18.9 cm **17** 3π cm **18** 144π cm^2 **19** 3.141 5927 **20** 15.71 m **21** 1885.7 cm^2 **22** 20 cm **23** 1608.5 cm^2 **24** 4 m **25** 39 cm **26 a** 6.7 cm **b** 35 cm^2 **27 a** 10 cm **b** 42.9 cm^2 **28 a** 8827 m^2 **b** 388 m **29 a** 1592.8 m^2 **b** 174.5 m
Page 86 **1** A **2** C **3** A **4** C **5** D **6** A **7** B **8** B **9** D **10** D **11** A **12** D **13** B **14** C **15** B
Page 87 **16** 6.283 19 **17** 28π cm **18** 81π cm^2 **19** 2.73 m **20** 2.72 m^2 **21** 67 cm **22** 201.1 cm^2 (1 d.p.) **23** 23 cm **24** 361π cm^2 **25** 38 cm **26 a** 1539.4 cm^2 (1 d.p.) **b** 159.9 cm (1 d.p.) **27** 256.9 cm^2 (1 d.p.) **28 a** 6.8 m^2 (1 d.p.) **b** 15.6 m (1 d.p.) **29 a** 25.1 cm (1 d.p.) **b** 36.5 cm^2 (1 d.p.)

Volume

Page 88 **1** D **2** C **3** B **4** B **5** B **6** C **7** A **8** D **9** C **10** D **11** B **12** A **13** D **14** C **15** A
Page 89 **1** C **2** C **3** B **4** A **5** B **6** D **7** C **8** D **9** C **10** D **11** A **12** B **13** A **14** A **15** B
Page 90 **16** 1000 cm^3 **17** 2500 **18** 800 mL **19** 216 cm^3 **20** 15.6 **21** 60 000 000 **22** 2.16 m^3 **23** 904.8 cm^3 **24** 1 000 000 **25** 210 cm^3 **26 a** 24 000 mL **b** 24 000 cm^3 **c** 20 cm **27** 4.41 m^3 **28 a** 790 020 mm^3 **b** 790 cm^3 **29 a** 16.8 m^2 **b** 1.68 m^3 **30 a** 14 m^3 **b** 14 000 L
Page 91 **1** B **2** D **3** D **4** A **5** A **6** D **7** B **8** D **9** B **10** C **11** C **12** C **13** B **14** A **15** A
Page 92 **16** 10 250 **17** 0.02 m^3 **18** 420 cm^3 **19** 240 mL **20** 16 cm^2 **21** 0.3 **22** 407 cm^3 **23** 30 762 cm^3 **24** 43 562 cm^3 **25** 900 **26 a** 138.6 m^3 **b** 138 600 **27** 1.365 m^3 **28 a** 87 965 cm^3 **b** 88 L **29** 33 **30 a** 80 cm **b** 2.304 m^3
Page 93 **1** B **2** C **3** A **4** C **5** D **6** B **7** B **8** D **9** A **10** A **11** B **12** D **13** C **14** D **15** C
Page 94 **16** 20 cm **17** 50 **18** 2000 L **19** 0.02 cm^3 **20** 15 m^3 **21** 3762 cm^3 **22** 0 (They have the same volume.) **23** 3 cm **24** 7741.8 cm^3 **25** 40 560 m^3 **26 a** 9315 cm^3 **b** 0.009 315 m^3 **27 a** 164 600 L **b** 15 000 L **28** 141 960 cm^3 **29** 8000 L **30** \$138

Time

Page 95 **1** B **2** C **3** D **4** A **5** D **6** B **7** B **8** C **9** D **10** A **11** B **12** D **13** A **14** C **15** C
Page 96 **1** D **2** B **3** C **4** A **5** D **6** C **7** D **8** A **9** A **10** B **11** B **12** C **13** D **14** C **15** B
Page 97 **16** 3 min 30 s **17** 225 **18** 2 h 50 min **19** 1945 **20** 8 h **21** 5 September **22** 12:05 pm **23** 92 **24** 8 am **25** 57 min 20 s **26 a** 1745 **b** 1 h 50 min **c** 9:52 pm **27 a** 1:16 pm **b** 3 h 20 min **c** 16 min **d** 11:53 **e** 10:52 **f** 1 h 29 min **g** He should get to the station at the same time the train is due to depart so he might or might not be able to catch the train. **28 a** 5 h 24 min **b** 2:42 pm

Page 98 **1** C **2** D **3** D **4** A **5** C **6** C **7** C **8** B **9** B **10** D **11** A **12** B **13** B **14** D **15** A
Page 99 **16** 7 h 15 min **17** 1 h 30 min **18** 2:15 pm Tuesday **19** 37 **20** 20 December **21** 18 h 50 min **22** 4:30 pm Friday **23** 5:45 pm **24** 9000 **25** 1 d 13 h 20 min **26 a** 8 h 20 min **b** 25 min **c** 12:30 pm **27 a** 19 August **b** Friday **c** 5:30 am **d** 2 years 9 months **28 a** 1730 **b** *Messy Moments* **c** 12:10 pm **d** 20 min **e** 22 **f** 58 min
Page 100 **1** B **2** C **3** A **4** D **5** B **6** A **7** C **8** C **9** D **10** A **11** B **12** C **13** D **14** B **15** A
Page 101 **16** 24 810 **17** 4 h 21 min **18** 9 months **19** 1 October **20** 17 min **21** 44 h 45 min **22** 1:15 pm 4 April **23** 1:23 pm **24** 2 d 6 h 25 min **25** 10:45 pm **26 a** 1 h 15 min **b** 4:50 pm **27** No, 3.91 s faster **28 a** 1 h 55 min **b** 6:30 pm **c** 1:10 am Wednesday **d** 5:10 pm Tuesday **e** 5:30 pm Thursday **29 a** 15 **b** 3:15 pm Thursday

Mixed Measurement

Page 102 **1** B **2** D **3** D **4** A **5** C **6** B **7** C **8** A **9** B **10** C **11** B **12** D **13** D **14** C **15** A
Page 103 **16** 320 000 **17** 54 cm^2 **18** 32 cm **19** 465 **20** 750 mL **21** 5π cm **22** 280 cm^3 **23** 15 August **24** 154 cm^2 **25** 8 000 000 **26** 5.04 ha **27** 2210 **28 a** 60 cm **b** 0.273 m^2 **c** 0.205 m^3 **29 a** 10 000 L **b** 12.5 L/min **30 a** 1.8 m **b** 2 m^2
Page 104 **1** B **2** C **3** A **4** B **5** C **6** B **7** A **8** B **9** C **10** A **11** D **12** C **13** D **14** A **15** D
Page 105 **16** 25 cm^2 **17** 121π cm^2 **18** 6000 L **19** 100 **20** 264 **21** 0.56 m^2 **22** 705 m^3 **23** 176 cm **24** 200 000 **25** 31 h 40 min **26 a** 3 m **b** 58.07 m^2 **c** 8.7 m^3 **27 a** 0.264 m^2 **b** 3.65 m^2 **28 a** 75 000 m^2 **b** 126 000 m^2 **c** 20.1 ha **d** 2.24 km **e** 4.8 km/h
Page 106 **1** D **2** C **3** C **4** B **5** D **6** D **7** B **8** B **9** C **10** B **11** A **12** A **13** A **14** D **15** C
Page 107 **16** 908 cm^2 **17** 80 000 **18** 400 ha **19** 120 cm^3 **20** 0.1995 m^2 **21** 2:04 pm **22** 3.04 L **23** 13 cm **24** 900 **25** 2197 cm^3 **26 a** $4056 **b** 87 360 **c** 4 d 1 h 4 min **d** $5724 **27 a** 0.624 m^2 **b** 70 cm **28 a** 962 cm^2 **b** 70π cm

Pythagoras' Theorem

Page 108 **1** C **2** D **3** B **4** D **5** B **6** A **7** D **8** D **9** A **10** C **11** C **12** A **13** B **14** A **15** B
Page 109 **1** C **2** B **3** A **4** B **5** A **6** C **7** A **8** D **9** C **10** D **11** D **12** A **13** C **14** B **15** B
Page 110 **16** 621 **17** 5.92 **18** rational **19** 89 **20** true **21** true **22** 24.6 cm **23** 11 **24** 51 **25** 44 **26 a** 38 cm **b** 19 cm **27** 616 m **28** 14.70 m **29 a** (below) **b** 3.3 m **30 a** 60 cm **b** 1.09 m

29 a

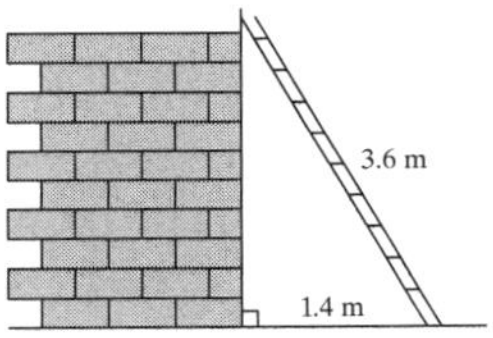

Page 111 **1** D **2** A **3** B **4** B **5** D **6** A **7** C **8** A **9** C **10** D **11** C **12** B **13** D **14** C **15** A
Page 112 **16** 93 **17** irrational **18** 87 **19** true **20** true **21** 169 **22** 15 **23** 34 cm **24** 7 **25** 84 cm **26 a** 14.9 m **b** 9.4 m **c** 35.3 m **27** 504 m^2 **28 a** 1.69 m **b** 5070 cm^2 **c** 60 cm **d** 25 cm **29** 160 cm
Page 113 **1** B **2** C **3** A **4** D **5** B **6** C **7** A **8** B **9** A **10** C **11** B **12** A **13** D **14** C **15** D
Page 114 **16** irrational **17** 37 **18** 9.70 **19** 78 **20** 39 **21** $\sqrt{39}$ **22** $\sqrt{58}$ **23** 565 **24** 12.5 **25** 4.1 **26 a** 37 249 m^2 **b** 456 m **27** 10.30 m **28** 276.06 m^2 **29 a** 40.9 **b** No, $AB^2 + BC^2 \neq AC^2$ **30** 53.2 m

Congruence and Properties of Quadrilaterals

Page 115 **1** A **2** B **3** B **4** D **5** C **6** B **7** A **8** B **9** D **10** C **11** A **12** C **13** D **14** A **15** C
Page 116 **1** D **2** D **3** A **4** C **5** D **6** A **7** A **8** B **9** B **10** A **11** C **12** C **13** B **14** C **15** B
Page 117 **16** equal **17** false **18** false **19** AAS **20** *RP* **21** 60° **22** true **23** false **24** *DE* **25** square **26 a** parallelogram **b** alternate, $PQ \parallel SR$ **c** alternate, $QR \parallel PS$ **d** common side **e** AAS **f** corresponding angles of congruent triangles **g** Opposite angles of a parallelogram are equal. **27 a** SSS **b** $\angle CDB$ **c** SAS **d** 90°; $\angle AXD = \angle CXD$ corresponding angles of congruent triangles, but *AC* is a straight line so $\angle AXD + \angle CXD = 180°$ **28 a** rhombus **b** 106°; diagonals bisect angles so $\angle ABC = 74°$, $\angle BAD + \angle ABC = 180°$, co-interior $AD \parallel BC$
Page 118 **1** A **2** A **3** C **4** B **5** D **6** B **7** D **8** B **9** A **10** C **11** D **12** C **13** D **14** C **15** B
Page 119 **16** equal **17** RHS **18** 70°, 110°, 110° **19** false **20** SAS **21** *CDA* **22** parallelogram **23** 80° **24** rectangle **25** rhombus **26 a** SAS **b** corresponding sides of congruent triangles **c** 38°, angle sum of triangle is 180° **d** ΔADP, ΔCDP **e** rhombus **f** corresponding angles of congruent triangles are equal **27 a** *MH* **b** 100° **c** reflection in *LM* **28 a** 30° **b** $PS = PQ$, sides of a square, $PT = PQ$, sides of equilateral triangle, so $PS = PT$ **c** 75° **d** obtuse-angled, isosceles

Page 120 1 C 2 A 3 B 4 B 5 C 6 C 7 A 8 D 9 A 10 B 11 C 12 A 13 D 14 D 15 D

Page 121 **16** squares, octagons **17** 135° **18** 40° **19** rectangle **20** 52° **21** 36° **22** *RSP* **23** rotation of 180° about *T* **24** 116° **25** false **26** Not necessarily. They are equiangular but need to have a pair of corresponding sides equal to be congruent. **27 a** $\angle QPR = \angle SRP$, alternate angles $PQ \parallel SR$ **b** $\Delta PQR \equiv \Delta RSP$ AAS **c** $\angle SPR = \angle QRP$ corresponding angles of congruent triangles, so the alternate angles are equal **28** 105°, $\angle PLN = 90°$ (angle of square) so $\angle OLN = 45°$ (diagonals bisect angles), $\angle NLM = 60°$ (angle of equilateral triangle) **29 a** RHS **b** isosceles, $AE = CE$ (corresponding sides of congruent triangles) **c** $BE \perp AC$, RHS; $AB = CB$, SSS; $\angle AEB = \angle CEB$, SAS

Probability

Page 122 1 B 2 D 3 C 4 B 5 C 6 D 7 A 8 A 9 C 10 B 11 C 12 D 13 B 14 A

Page 123 1 B 2 A 3 D 4 C 5 D 6 A 7 B 8 A 9 A 10 B 11 C 12 B 13 D 14 D 15 C

Page 124 **16** $\frac{999}{1000}$ **17** $\frac{5}{6}$ **18** $\frac{14}{23}$ **19** $\frac{1}{52}$ **20** $\frac{4}{13}$ **21** one **22** 10 **23** $\frac{5}{7}$ **24** false **25** $\frac{1}{2}$ **26 a** $\frac{1}{4}$ **b** $\frac{1}{12}$ **c** $\frac{1}{3}$ **d** $\frac{7}{12}$ **27 a** 9 **b** 6 **c** 33 **d** 47 **e** $\frac{16}{33}$ **f** $\frac{24}{35}$ **28 a** 36 **b** $\frac{13}{25}$ **c** $\frac{16}{43}$ **d** $\frac{4}{13}$ **e** $\frac{37}{100}$

Page 125 1 D 2 C 3 D 4 A 5 C 6 A 7 C 8 A 9 A 10 C 11 B 12 D 13 B 14 B 15 D

Page 126 **16** $\frac{5}{6}$ **17** $\frac{1}{3}$ **18** $\frac{5}{6}$ **19** $\frac{2}{3}$ **20** true **21** $\frac{1}{52}$ **22** $\frac{7}{12}$ **23** $\frac{1}{5}$ **24** 5 **25** $\frac{5}{12}$ **26 a** 11 **b** 4 **c** $\frac{5}{12}$ **d** $\frac{13}{30}$ **e** $\frac{6}{13}$ **f** $\frac{1}{3}$ **27 a** (see table below) **b** $\frac{39}{194}$ **c** $\frac{78}{223}$ **d** $\frac{13}{25}$ **28 a** (see diagram below) **b** $\frac{4}{23}$ **c** $\frac{13}{23}$

27 a

	Junior	Senior	Total
Male	78	145	223
Female	72	93	165
Total	150	238	388

28 a

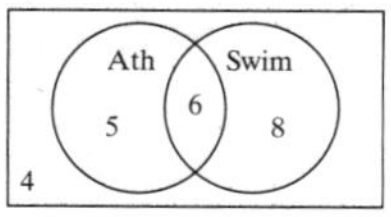

Page 127 1 D 2 D 3 C 4 C 5 D 6 A 7 C 8 D 9 B 10 A 11 C 12 B 13 B 14 B 15 A

Page 128 **16** $\frac{5}{6}$ **17** $\frac{2}{13}$ **18** $\frac{7}{25}$ **19** $\frac{1}{10}$ **20** $\frac{3}{20}$ **21** rolling 1 **22** $\frac{2}{3}$ **23** $\frac{2}{9}$ **24** $\frac{4}{9}$ **25** $\frac{10}{13}$ **26 a** 16 **b** $\frac{3}{50}$ **c** 73 **d** 71 **e** $\frac{17}{25}$ **f** $\frac{11}{32}$ **g** $\frac{51}{73}$ **27** 16 **28 a** 75 **b** (see below) **c** $\frac{31}{75}$ **d** $\frac{5}{14}$ **e** 473

28 b

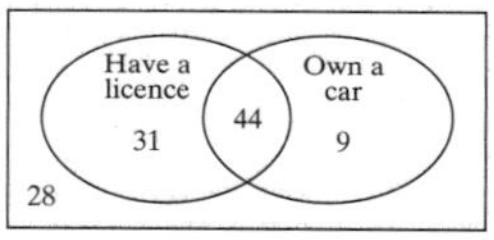

Statistics

Page 129 1 D 2 D 3 C 4 B 5 A 6 C 7 B 8 A 9 B 10 B 11 D 12 C 13 C 14 D 15 A

Page 130 1 A 2 B 3 D 4 D 5 A 6 A 7 C 8 D 9 D 10 A 11 C 12 C 13 C 14 B 15 B

Page 131 **16** 5 **17** 6.5 **18** $6\frac{5}{6}$ **19** 7 **20** 42 **21** true **22** 36 **23** 106 **24** 100 **25** $101\frac{5}{18}$ **26 a** It means that every student in the school was surveyed. **b** discrete numerical **c** 689 **d** 569 **e** 0 **f** 0.8 **27 a** continuous numerical **b** 2096 g **c** 1891 g **d** 3273 g **e** 3285.5 g **f** range 1083 g, mean 3380 g, median 3357 g; the range decreases, while the mean and median increase **g** If it is unusual to have a baby of this birth weight, then leaving it out will give a truer picture of the size of babies normally born at the hospital. But if it is quite common to have a baby of that birth weight, then a more accurate picture is given when it is included.

Page 132 1 A 2 C 3 A 4 D 5 C 6 B 7 A 8 B 9 B 10 D 11 C 12 A 13 C 14 D 15 B

Page 133 **16** 16 **17** 1 **18** 33 **19** true **20** true **21** 25 **22** categorical **23** 77 **24** white **25** $\frac{1}{7}$ **26 a** A census surveys the whole population while a sample surveys part of the population. **b** It is easier. If the population is large it might be very difficult to conduct a census. **27 a** 38 min **b** The spread is quite large with a range of 45 minutes but most of the times are between 34 and 43. There is a gap between 43 and 51 and then another smaller cluster. There is another gap between 60 and 66 and between 69 and 79. **c** 39.5 min **d** 45.6 min (1 d.p.) **e** The median gives a better idea of the usual time to drive to work. The mean is affected more than the median by the few longer times. **f** 79 is an outlier so could be ignored. It would depend on how frequently a time like that occurs whether a truer picture of the times to travel to work would be found with or without that time. **28 a** 80.75 cm **b** 85 cm **c** 13 cm **d** This sample of workmen is not likely to be representative of those who will be using the park. The sample should include women and children.

Page 134 1 A 2 D 3 A 4 C 5 B 6 D 7 D 8 C 9 C 10 B 11 A 12 A 13 B 14 B 15 D

Page 135 **16** false **17** 125 **18** 19 **19** false **20** 0.75 **21** true **22** false **23** $\frac{341}{480}$ **24** $\frac{17}{24}$ **25** $\frac{7}{40}$ **26** 20

27 No information is given about the survey so no conclusions can be drawn about it. The sample might have been very small and unrepresentative of the general population. The results could be biased. **28 a** 19 **b** (see below) **c** 27 **d** 28 **e** There is a cluster of ages from 23 to 28 and another smaller cluster around 37. The age of 39 might be classed as an outlier, or all four ages between 36 and 39. **29 a** Aaron found the mean of the two means, but because the two samples are of different sizes, this is not the combined mean. The correct mean is $18.4\dot{3}$. **b** There is not enough information to determine whether the difference between the means was too large. Aaron should conduct further samples, conduct a census, or study the samples to see if an outlier could be ignored to produce a more consistent result.

28 b

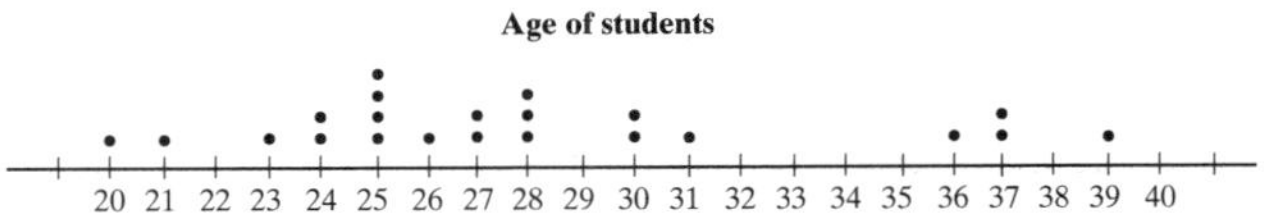

Exam Paper 1

Page 136 1 B 2 D 3 B 4 A 5 A 6 C 7 D 8 B 9 C 10 C 11 D 12 D 13 D 14 A 15 A 16 B 17 B 18 C 19 C 20 A 21 B 22 A 23 C 24 D 25 A

Page 138 **1** 1 **2** –5 **3** $0.\dot{6}$ **4** $2(x-4)$ **5** \$900 **6** 14 cm **7** 10 **8** 32 cm **9** 6:3:2 **10** $10a^2$ **11** 70 km/h **12** 8 **13** $x = 6$ **14** $2x^2 + 10x$ **15** –10 **16** false **17** $\frac{3}{8}$ **18** 5^5 **19** 10% **20** rectangle (or square) **21** 2 pm **22** 0.263 **23** $-6a + 15b$ **24** 8 **25** 6 **26** AAS **27** $\frac{8}{15}$ **28** 33 cm² **29** $y = x + 7$ **30** 80 cm³

Page 140 **1 a** –1 **b** –40 **2 a** 2:35 pm **b** $2\frac{1}{2}$ h **c** 200 km/h **3** $x = 60$ **4 a** 10 m **b** 1:40 **5 a** $11x - 15y$ **b** $n^2 + 4n - 10$ **6 a** \$3000 **b** 25% **c** \$12 000 **7 a** $y = x - 2$ **b** 2, 4, 6, 8 **c** (see below) **d** (–4, –6) **8 a** 3 **b** 8 **c** 7.5 **d** 7 **e** $\frac{7}{10}$ **f** No difference to the mean, mode or range. The median will increase to 8. **9** 3^{12} **10 a** 0.625 **b** $0.\dot{4}$ **c** rational, they can be expressed as fractions **11 a** 56 cm **b** 8 **c** 144 cm²

7 c

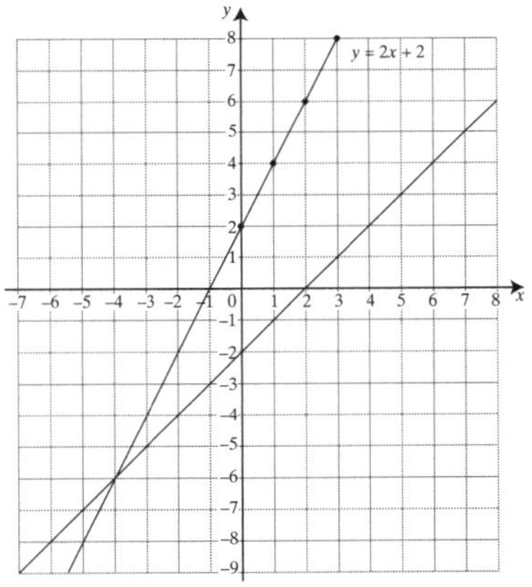

Exam Paper 2

Page 142 1 A 2 B 3 C 4 A 5 A 6 A 7 D 8 A 9 C 10 D 11 B 12 C 13 D 14 A 15 A 16 D 17 B 18 C 19 D 20 D 21 C 22 B 23 B 24 B 25 C

Page 144 **1** 0 **2** 40 **3** 1 000 000 **4** 7^6 **5** $24a^2b$ **6** 250% **7** $7(xy - 1)$ **8** $\frac{1}{2}$ **9** irrational **10** 7:200 **11** 80 **12** 5^{10} **13** 7:35 pm **14** 15 **15** 5 **16** $-2x^2 + 10x$ **17** 21 **18** $-\frac{1}{6}$ **19** \$240 **20** 0.289 **21** $x = 20$ **22** 170 **23** $\frac{4a}{b}$ **24** 8 **25** 600 cm² **26** 75 **27** $\frac{3}{8}$ **28** $3x(2x - 3)$ **29** 3^{10} **30** 80 km/h

Page 146 **1 a** 68 **b** –2482 **2 a** $2x + 27y$ **b** $x^2 - 9x - 14$ **3 a** 15 **b** $\frac{11}{20}$ **c** 40 **4 a** $x = 3$ **b** $x = 9$ **5** $0.\dot{2}\dot{7}$ **6 a** \$320 **b** \$352 **c** \$84/h **7 a** census **b** categorical **c** 25% **d** 99° **e** $\frac{7}{20}$ **f** 6 **8** 25% **9** 2^{14} **10 a** 35 **b** 3:1 **11 a** 14 **b** 4 **c** 4.5 **d** 6 **e** $4\frac{7}{9}$ **f** The range decreases to 5, the mode doesn't change and the median decreases to 4. **12 a** trapezium, the top and bottom sides are both perpendicular to one of the sides so they are parallel **b** 0.8 ha **c** 380 m

Exam Paper 3

Page 148 **1** A **2** A **3** B **4** A **5** C **6** B **7** D **8** B **9** A **10** D **11** B **12** C **13** C **14** C **15** B **16** D **17** C **18** B **19** A **20** C **21** D **22** D **23** D **24** B **25** A

Page 150 **1** –170 **2** 70 000 **3** 23 November **4** 36 **5** $-15xy + 3x^2 + 3x$ **6** 200% **7** 1 : 40 **8** $\frac{6a}{c}$ **9** 12 cm, 8 cm, 8 cm **10** 2^7 **11** –20 **12** 67.5 km **13** 5 **14** repeat **15** –1 **16** \$140 **17** $x = 3.5$ **18** $\frac{1}{4}$ **19** $-4q(2p + 1)$ **20** –1 **21** 20° **22** 3700 **23** 5^{12} **24** $y = 3x - 5$ **25** 12π cm³ **26** $-9p^2 + 12p + 11q$ **27** 500 000 **28** 0.1 L **29** 0.17 **30** \$1200

Page 152 **1 a** –6930 **b** 7 **2 a** \$7.50 **b** \$27.30 **c** \$15 **3 a** It can be expressed as a fraction. **b** $2.\dot{6}\dot{3}$ **4 a** (see below) **b** $\frac{11}{40}$ **c** $\frac{29}{40}$ **5 a** 2^6 **b** 3^{10} **6 a** alternate, $PQ \parallel SR$ **b** PR is a common side; SAS **c** The quadrilateral is a parallelogram because one pair of opposite sides are equal and parallel and corresponding angles of congruent triangles are equal. **7 a** $x = -1.5$ **b** $x = -8\frac{1}{3}$ **8 a** 120 g/L **b** 900 g **c** 11 **9 a** $\frac{4}{15}$ **b** 6 : 4 : 5 **c** 120 white, 100 green **10** 83.75 **11** 38.5 cm² **12 a** 24 cm² **b** The diagonals of a rhombus bisect each other at right angles. **c** 5 cm **d** 4.8 cm

4 a

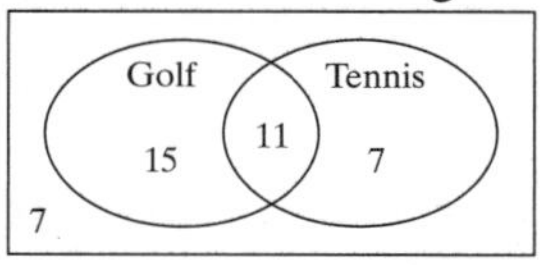

Exam Paper 4

Page 154 **1** B **2** A **3** D **4** D **5** A **6** D **7** D **8** B **9** B **10** A **11** D **12** C **13** C **14** B **15** A **16** A **17** B **18** C **19** D **20** C **21** A **22** C **23** B **24** C **25** A

Page 156 **1** 9 **2** $\frac{5}{6}$ **3** $9(3a + 5b)$ **4** 153 cm² **5** 6.5 **6** $7x^2 + 8x$ **7** –60 **8** 8 **9** 6 : 8 : 9 **10** –15 **11** \$1275 **12** $x = 5$ **13** $-\frac{3}{7}$ **14** 18 **15** $21x - 56$ **16** $7m$ **17** 0.1875 **18** \$340 000 **19** 1.7 m **20** $18a^2$ **21** $1\frac{5}{6}$ **22** 9 L/min **23** (1, 0) **24** YZ **25** $\frac{23}{90}$ **26** 7^6 **27** 52.54 m³ **28** \$588 **29** 6 cm **30** 5

Page 158 **1** $-19x - 16y$ **2** $5a^2b(3b - 2)$ **3 a** 375 m **b** 22 500 m **c** 22.5 km/h **4 a** rational, all fractions are rational **b** $0.\dot{4}2857\dot{1}$ **c** 0.429 **5 a** 153.94 cm² **b** 196 cm² **c** 42 cm² **6 a** 111 **b** $\frac{2}{15}$ **c** $\frac{41}{90}$ **d** $\frac{53}{90}$ **7 a** $x = 1.6$ **b** $x = -29$ **8 a** \$2250 **b** 12% **c** \$2628 **9 a** 7, 5, 3, 1 **b** (see below) **c** $x = 8$ **10 a** 19 **b** 24 **c** 29 **d** 30 **e** 30. The score being added is equal to the mean so the mean will not change. **f** It is not possible to tell without further information whether or not the group is representative. Because it is only a small sample and the students are all enrolled in the same course it is possible that it is not representative. **11 a** 11 cm **b** 330 cm² **c** 6600 cm³ **d** 6.6 L **e** 1.4 L

9 b

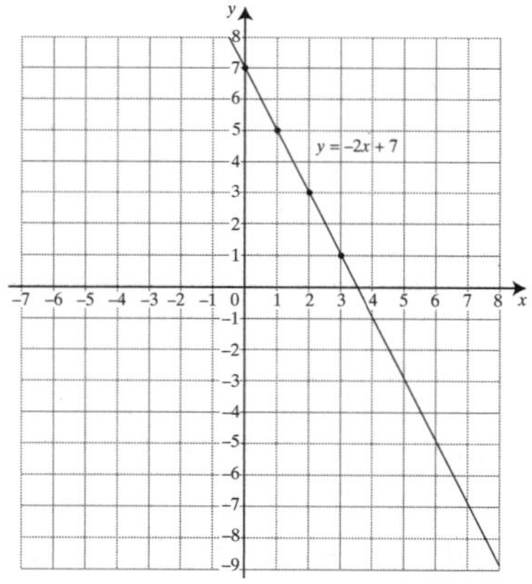

Exam Paper 5

Page 160 **1** A **2** D **3** B **4** C **5** B **6** B **7** D **8** D **9** B **10** C **11** D **12** C **13** A **14** A **15** B **16** C **17** A **18** A **19** C **20** C **21** B **22** D **23** A **24** C **25** D

Page 162 **1** $96a^2$ **2** 0.47 **3** –6480 **4** 25 m² **5** 1 : 5 **6** 100% **7** $-8\frac{1}{8}$ **8** $x = 8$ **9** $-6x^2 + 15x$ **10** 4.8 m³ **11** $2\frac{3}{4}$ h **12** $4xy(2x - 1)$ **13** $-8x^2 + 11x$ **14** 2^{12} **15** 9 **16** 3 h 15 min **17** (0, 4) **18** –24 **19** 87.125 **20** $\frac{99\,999}{100\,000}$ **21** 80° **22** square **23** 973 **24** 268 **25** $133\frac{1}{3}$% **26** 3:25 pm **27** 35 831 809 **28** –16 **29** 30 cm **30** 300

Page 164 **1** $-9x + 2$ **2 a** \$704 **b** \$80 **3 a** 36, 44, 52 **b** To find the number of matches multiply the stage number by eight and add four. **c** $m = 8s + 4$ **d** 140 **e** 29 **4** $x = 3\frac{6}{11}$ **5 a** numerical **b** A census would use the heights of all the students in the school.

c 34 cm **d** $165\frac{2}{3}$ cm **e** 167 cm **f** 166.5 cm **g** None increase. The mean, median and range would all decrease. **6 a** 42 m^3 **b** 42 000 L **c** 5 L/min **7 a** 100 **b** $\frac{2}{5}$ **c** $\frac{59}{100}$ **8 a** Opposite sides of a parallelogram are equal. **b** vertically opposite angles **c** alternate, $AD \parallel BC$ **d** congruent, AAS **e** corresponding sides of congruent triangles **f** The diagonals bisect each other. **9** 1 **10 a** 15.3 m **b** 171.5 m^2 **c** 53 m

Exam Paper 6

Page 166 1 C 2 C 3 A 4 B 5 D 6 C 7 A 8 B 9 D 10 D 11 C 12 C 13 C 14 A 15 B 16 A 17 B 18 B 19 A 20 B 21 D 22 D 23 A 24 D 25 C

Page 168 **1** 22.0 **2** 3^{20} **3** $133\frac{1}{3}\%$ **4** $2m$ **5** 9 cm **6** categorical **7** $5\frac{1}{2}$ **8** 24 **9** $\frac{18}{35}$ **10** 25 **11** 216 cm^2 **12** $-7x^2 - 15x$ **13** 2.22 pm **14** $0.\dot{3}0769\dot{2}$ **15** $3mn(3m - 2n)$ **16** $\frac{999\,999}{1\,000\,000}$ **17** 58°, 122°, 122° **18** $4\frac{2}{3}$ **19** 93 **20** $x = 2.8$ **21** $\sqrt{77}$ **22** \$1280 **23** $-15x^2 + 6xy - 3xz$ **24** 3.6 km **25** $y = 12x - 8$ **26** 4 h 12 min **27** $\frac{8}{33}$ **28** 8.5 cm **29** 1 440 000 **30** 35 520 cm^3

Page 170 **1** $9p^2 - 12p - 68$ **2 a** 195 **b** 14 **c** $\frac{1}{13}$ **d** $\frac{18}{25}$ **3 a** 0.75% **b** \$967.20 **c** \$2020 **4 a** It doesn't form a straight line. **b** (1, 0) **c** –4 **d** 4 **e** $y = 4x - 4$ **f** $y = 4x$ **5 a** $7\frac{23}{24}$ **b** $7.958\dot{3}$ **6 a** 460 m **b** 9 min 35 s **7 a** $8(x - 13) = 3x + 11$ **b** 23 **8 a** reflection in *AC* **b** SSS **c** Corresponding angles of congruent triangles are equal. **d** One pair of opposite angles is equal. **e** rhombus **9 a** 4 h 35 min **b** 80 km/h **c** 25.5 L **10** 67.2 L **11 a** –5 °C **b** –2 °C **c** The range is 16° but –13 °C is an outlier and if that is ignored the range is 10°. Most of the temperatures fall between –5 °C and 1 °C. **d** $-2\frac{1}{3}$ °C **12 a** 24 cm **b** 288 cm^2 **c** Diagonals of a square are equal and bisect each other at right angles. **d** 17.0 cm **e** 144π cm^2 **f** 164 cm^2

NOTES